# PROJECT MINDSHIFT

# PROJECT MINDSHIFT
# PROJECT MINDSHIFT
# PROJECT MINDSHIFT
# PROJECT MINDSHIFT
# PROJECT MINDSHIFT

## The Re-Education of the American Public Concerning Extraterrestrial Life, 1947–Present

Michael Mannion

M. Evans and Company, Inc.
New York

M 38738839

M. Evans and Company, Inc.
216 East 49th Street
New York, New York 10017


**Library of Congress Cataloging-in-Publication Data**

Mannion, Michael T.
   Project mindshift : the re-education of the American public concerning extraterrestrial life, 1947 to the present / Michael Mannion.
      p.   cm.
   ISBN 0-87131-856-3 (hardcover)
   1. Human-alien   encounters.   2. Alien   abduction.   3. Unidentified flying objects.   I. Title.
   BF2050.M35   1998
   001.942—dc21                                          98-6510

DESIGN AND TYPESETTING BY RIK LAIN SCHELL

Manufactured in the United States of America

9 8 7 6 5 4 3 2 1

Photo credits for page 96: Photo of Dr. Edgar Mitchell courtesy Dr. Edgar Mitchell. Photo of Stanton Friedman courtesy Stanton Friedman. Photo of Budd Hopkins courtesy Budd Hopkins. Photo of Dr. John E. Mack courtesy Dr. John E. Mack. Photo of Raymond Fowler courtesy Raymond Fowler. Photo of Dr. David Jacobs courtesy Dr. David Jacobs. Photo of Don Berliner courtesy Don Berliner. Photo of Michael Lindemann courtesy Michael Lindemann.

To Trish, with whom I saw the light

# CONTENTS

*Life is not as it appears;*
*Neither is it otherwise.*

—Zen koan

*no longer dreaming*
*neither butterfly nor man*
*energy streaming*

—Michael Mannion
August 1997
Rangeley, Maine

# FOREWORD

Are we alone in the universe? If not, are there life forms more highly evolved than us? If so, have they contacted us? And if that is so, have they intervened in human affairs? These questions are the essence of what ufology, the scientific study of the UFO phenomenon, attempts to answer.

In this fascinating book, Michael Mannion looks at that variety of influences—some obvious, some subtle, even covert—which have shaped the predominant images of UFOs in the public's mind. His Mindshift Hypothesis is: Our world has been, and is now being, visited by advanced intelligent entities "from elsewhere" and this reality has been known to a limited number of people within the U.S. Government since at least 1947.

If you are a disbeliever about the UFO phenomenon but you follow the data and logic of Mannion's presentation, you will be led inexorably to jettison your disbelief. You may not become a believer, but you will surely be more open-minded because the weight of evidence and argument demand it from anyone with intellectual honesty.

If you are a skeptic, you will be shown that there is a strong case for the reality of the UFO phenomenon—stronger than you may have been willing to allow.

If you are a believer, you will have your beliefs tested for validity because the field of ufology is one in which many experts who accept the reality of the UFO phenomenon nevertheless disagree about many aspects of the phenomenon they study.

What exactly is a UFO? Here is the definition used by the Mutual UFO Network (MUFON), the largest public UFO research group in the world:

UFOs are objects observed in the skies or on the surface of the Earth which defy conventional explanation after a thorough study and investigation by competent people. Some of the things reported as UFOs are balloons, planets, meteors, satellites, stars, advertising aircraft, falling aerospace debris, and the like. These are IFOs—identified flying objects. The "hard" sightings which are yet to be explained are daylight discs, objects with unusual lights which are simultaneously tracked visually and on radar at fantastic speeds, objects which leave physical evidence after landing, authenticated photographs, and an increasing number of cases involving visitations or close encounters by humanoids or entities. It is this category which serious ufologists are trying to gather additional evidence in our quest to resolve the UFO phenomenon.

With that definition as the basis for systematic research, it is now abundantly clear after half a century of observations and data-gathering that UFOs are real. Here's my view of the situation, as someone who has been actively engaged in exploring it for several decades.

First, it appears that Earth is one of the major crossroads of local space. Consider just the Humanoid Catalog data base of mo٠ ٠n 3,000 UFO cases in which human-like entities were reported. Analysis of the data reveals a dozen types of humanoid lifeforms having two arms, two legs, a trunk and a head. Their size ranges from a mere foot in height to twelve feet or more. Beyond the HumCat data base are reports of UFO occupants who resemble reptiles more than humans, and reports of others who are more like insects than humans. Still others describe beings so exotic that they have no specific form. Shades of *Men in Black*!

Second, the Ancient Astronaut hypothesis, which researchers have been investigating for more than three decades, suggests that UFOs and their occupants have been in contact with humanity for millennia. Sometimes those ancient contacts were low-profile; at other times the beings seemed to intervene quite directly in the affairs of primitive peoples, giving rise to the notion of "gods from outer space" which has been recorded in cultures around the world. Have we humans been genetically engineered, as some researchers claim? Have our belief systems been manipulated by alien beings for obscure purposes, as other researchers claim? Shades of *2001*!

The evidence for this, I acknowledge, is more circumstantial and inferential than hard and physical. Considered objectively, however, there is nothing inherently implausible about the notion that more highly evolved life forms exist and have contacted us. Exobiology, that branch of science which searches for extraterrestrial intelligence, suggests that life will arise wherever conditions are not simply favorable but merely a little better than totally hostile. Since our sun is a relatively young star, there are many older star systems where organisms probably developed earlier than here on Earth and have evolved technological civilizations capable of space travel. So quite apart from ufology, it makes sense to say we are not alone. That is why there is a mainstream science program called SETI, or Search for ExtraTerrestrial Intelligence, which seeks contact with alien societies by using radiotelescopes to listen for signals from space.

The evidence from ufology ties in neatly with exobiology. Yet for all that evidence, mainstream science resists the idea of UFOs. The scientific community, by and large, does not accept ufology as a colleague discipline. Why? Dr. Michael Swords of Western Michigan University published an excellent paper in the 1989 *Journal of UFO Studies* demonstrating irrefutably that the ET hypothesis which the SETI scientists are pursuing is perfectly applicable for use in evaluating UFO reports. So the unwillingness of the scientific community to listen to ufologists is simply illogical and unscientific. The rejection really is due to nothing more than irrational prejudice.

Recognizing that irrationality segues us to the theme of Michael Mannion's book, *Project Mindshift.* Mannion describes how and why public opinion—the mind of the public, if you will—is shifting away from the disbelief characteristic of the scientific community and in general seems to be ahead of the scientific community on the UFO learning curve.

No longer is the UFO experience conceived in simplistic terms of little green men in spaceships with Martian license plates. That stereotyped image from the 1950s and 1960s was, quite literally, ridiculous. Or to put it more precisely, despite the facts which provide the basis for that stereotype, ridicule tended to greet the small number of people who professed belief in life on other planets which was visiting Earth.

The idea of ET visitation had been recognized well before Orson Welles' 1938 radio program, "The War of the Worlds," frightened a lot of East Coast radio listeners with his tale about Martian invaders. Throughout the end of the nineteenth century, there had been a great debate in astronomy about whether markings on Mars were canals indicative of a civilization there. In 1898, H. G. Wells wrote his novel *The War of the Worlds* on the premise that Mars was inhabited by an advanced civilization hostile to Earth; Orson Welles used it 40 years later as the basis of his radio program. In 1899, the renowned scientist-inventor Nikola Tesla had claimed (mistakenly, it now seems) to receive radio signals from space, probably Mars, and had gotten great public attention over it. So the idea of ETs in contact with Earth was a familiar one when ufology began in 1947 with media reports that June about Kenneth Arnold's sighting of nine saucer-shaped UFOs near Mt. Ranier and then in July about the now-famous incident of a crashed UFO retrieved by the Army Air Force near Roswell, New Mexico. But it was an idea akin to science fiction, and a variety of influences reinforced that perception.

Today, the UFO phenomenon is becoming less associated with science fiction and more associated with science. For example, serious scientists now ask such questions as: What exactly are UFOs? Where do they come from? How do they propel themselves? Are they physical or otherwise? What do they want, if anything? What do we know about their occupants? There is such a wide variation in their reported characteristics—such as size, shape, coloration, material, types of occupants, their behavior and so forth—that it is terribly confusing to the newcomer, especially when, as I noted above, many authorities disagree with one another about these and other questions.

In my judgment—for whatever it may be worth to you, the reader—no single explanation can cover all the experiences and events lumped together under the label "the UFO phenomenon." It would be nice if there were a neat, simple explanation, but I find there isn't. No, the UFO phenomenon is multileveled and can't be explained in a unified fashion. (Mannion, I'm pleased to say, concludes likewise.) As I survey the evidence, there are three "answers" to the UFO problem—three qualitatively different answers. Those answers relate to what can be called different levels of reality or different aspects of our very existence. Simply put, the levels or aspects are: terrestrial, extraterrestrial and metaterrestrial. (The latter term was coined by the late Dr. J. Allen Hynek to denote an unseen world beyond the familiar three-dimensional world we know. It is synonymous with metaphysical, meaning "beyond the physical," but it has a more scientific connotation to it, like hyperspace and interdimensional do.)

However, the details about all that are a subject for another day. For now, I'm pleased to offer this foreword to Mannion's excellent survey and discussion of the UFO phenomenon—its history, its data, its public perception, its covert manipulation by a U.S. military-intelligence group whose actions and motives should be examined publicly. Mannion has a strong grasp of the facts and a most engaging way of presenting them. No matter what your view of UFOs, you'll find this book is a "really good read."

**John White** is an author in the fields of consciousness research and higher human development. His fifteen books include *The Meeting of Science and Spirit* and *What Is Enlightenment?* His writing has appeared in the *New York Times*, *Reader's Digest*, *Esquire*, *Omni*, *Woman's Day*, and various other publications. He produces an annual conference, "The UFO Experience." He lives in Cheshire, Connecticut.

*Has a clandestine group formed a shadow government in America, seized control of all information concerning extraterrestrials and UFOs, and kept this knowledge from the highest political and military leaders in the country? One of only twelve men to have walked on the Moon believes this to be the case.*

*Are human beings the world over being abducted by extraterrestrial entities? An internationally renowned psychiatrist works with hundreds of people who claim to have been abducted by aliens—and believes their stories.*

*Do the intelligent astronauts from other solar systems exploring Earth have an agenda? A top nuclear physicist, formerly with the U.S. space program, asserts that the aliens do have an agenda—to prevent humanity from exploring space until humans become less violent.*

T he field of ufology—the scientific study of the UFO phenomenon—is filled with all sorts of stories and speculations, from the mundane to the miraculous. Tales are told by top scientists, military personnel, and world leaders, as well as by neighbors, friends, and family members. How is one to make sense of it all? How can one evaluate this vast body of knowledge and separate the truth from fantasy or fiction?

For anyone who has never seen a UFO—and especially for anyone who has seen a UFO—*intuition* may be more valuable than *information*.

Is intuition too subjective to be relied upon? So many personal variables come into play that it may seem difficult to form an opinion on complex subjects such as UFOs and extraterrestrial life based on intuition. However, intuition may not be as subjective as it seems at first glance.

Information, on the other hand, may seem initially to be more dependable. Information appears to be objective, independent of the individual receiving it. Basing an opinion on information may seem to be a more sound approach. However, naive reliance on objective information can sometimes lead to ruin.

In 1964, for example, the Johnson Administration announced to the nation that the North Vietnamese had launched an attack on American forces in the Gulf of Tonkin. The so-called "Gulf of Tonkin Incident" was used by President Johnson to pressure Congress into giving him unprecedented power to wage war without congressional approval. The American people trusted their leaders and supported the President *based on the information they received.* This support led to the Vietnam War, in which an estimated 55,000 Americans and 4,000,000 Southeast Asians died.

However, there was one major flaw in this process: the information the American public received was false. The Gulf of Tonkin incident never occurred. It was fabricated by the Johnson Administration to provide a needed pretext to go to war. In 1989, after twenty-five years of deception, the U.S. government publicly acknowledged that the Vietnam War had begun with false information. This was reported in mainstream publications such as the *New York Times.* The American public had been fed a deliberate lie by their government.

As this example tragically shows, information may not be as objective as it seems. Information must be verified because it may prove to be false. Incomplete information also can be misleading, either intentionally or unintentionally. The source of the information must be considered carefully when evaluating it. And information must be *interpreted.* Interpretation brings the subjective element back into play.

Intuition can be a powerful tool, and trust in one's intuition can lead to wise decisions. Human beings use it all the time in important situations. For example, intuition is crucial in deciding whether or not to marry someone. Intuition is also a critical factor in the life-and-death deliberations of members of a jury. Following one's intuition can often be the best course of action.

In the following pages, information from a wide range of

sources will be presented. Each reader will need to keep an open mind; each reader will need to be a true skeptic. Is an internationally renowned figure more believable than a four-year-old child? Is a UFO investigator with access to secret official government information a more reliable source than an ordinary person who reports an extraordinary experience? At times, the child may know more than the expert, the ordinary citizen more than the experienced UFO investigator.

Are intelligent beings from other worlds visiting Earth? Are UFOs the visible manifestation of their presence on our planet? The reader will have to listen to his or her inner voice and, based on the preponderance of the evidence, decide what has the ring of truth.

Stand beneath the stars at night and hold one hand up to the sky. Hundreds of millions of stars exist in the space blocked out by one hand. And hundreds of millions of planets as well. On some of those planets, life evolved billions of years before it did on Earth. We mastered flight less than one hundred years ago and now are exploring our solar system. What would intelligent beings with a billion-year head start on us be up to? Which is more preposterous—that other intelligent living beings are exploring this part of the cosmos, or that we are the only intelligent beings in all of Creation?

The UFO phenomenon presents a great challenge to mankind. It is a complex, multi-level phenomenon which is not yet understood. It is a mystery that forces anyone who considers it to throw out all preconceptions and look at the emerging facts about the cosmos and our place in it with a completely open mind. In this way, the UFO phenomenon may indirectly expand one's consciousness, enhance one's reverence for our planet, and deepen one's awe at the magic and majesty of the universe.

This is a book of questions, not answers. The solution is up to each reader. No one has an answer to date. Each researcher, investigator, or abductee has contributed only a piece of the puzzle. Here are some pieces of the puzzle picked up along the way, offered as a sampling of things to come.

## 1943

A young mother is alone in the bedroom with her two-year-old child. Her husband is working at night as part of the war effort. The streets are completely dark. The World War II blackout is in effect. Suddenly, the young woman is "frozen." Five "little men" come through the window into the bedroom. They gather around her young child. She can do nothing. After some time, they float back out through the window into the night.

Decades later, the woman reported that the "little men" were frequent visitors.

## 1960

A ten-year-old looked up from his hospital bed to see five small, thin figures standing beside his bed. They were as gray as the faint gray light that dimly illuminated the pediatric ward in which the youngster lay, recuperating from surgery. The child was heavily sedated and could not move. Hours later, inexplicably, the youngster was found wandering on a different floor in another wing of the hospital. None of the nurses saw the child leave the pediatric unit. No one on the staff could understand how the child could even be walking so soon after serious surgery.

## 1970

Two young couples were camping in an extremely remote section of mountainous country. They were isolated in a forest, miles from the nearest campsite. In the middle of the night, they all awoke to find themselves standing outside the tent, confused and frightened. Something disturbing had occurred but no one could say what it was. One of the young men suddenly began asking excitedly, "Did you see the bear? Did you see its eyes?" He grabbed a rifle they had brought for protection and began to roam the periphery of the campsite. No bear prints were evident; none of the food the inexperienced campers had left outside the tent had been touched. After awhile, they calmed down and finally slept.

One month later, one of the four had a vivid dream of being aboard an alien space ship. Many of the incidents in the dream began to appear in books—books published fifteen years later.

## 1975

A man in his twenties called his older brother at 3:30 in the morning. The two men lived in a major metropolitan area. He fearfully told his sibling that the aliens were nearby. "They're here! I just know it; they're here!" The two brothers stayed on the phone for an hour or so until the younger brother felt at ease. Both awoke to hear radio news reports that a UFO had been reported at 3:30 A.M. less than a mile from the young man's home. There were over a dozen eyewitnesses.

A few years later, a casual acquaintance said to the younger sibling, "I don't know why I can tell you this but I just feel that I can." He proceeded to give an eyewitness account of the UFO incident the young man had sensed was in progress.

## 1979

A man in his mid-thirties was just about to fall asleep when a strange shape entered his bedroom. He heard a telepathic voice say, "He's not ready." As the shape hung in the air, his right arm slowly raised itself and pointed upward at the mysterious object. The man's dog looked up from sleep, saw the object, bared its fangs, and leaped into the air. For nearly an hour, the dog barked and cried so loudly that neighbors threatened to call the police. Finally, the dog quieted down and the man returned to bed. Just as consciousness was slipping away, the shape returned to the room. He heard a telepathic voice say, "He's ready."

Two years later, he said to a relative, "The strangest thing came into my room." He told the story as if it had just happened. Many years later, when he first saw the Stealth Bomber, he recognized that the shape and the bomber were made from similar material.

## 1989

A four-year-old awoke and came to the bedside of a couple visiting his parents for the weekend at their new home. "Can I come into bed with you?" the child asked plaintively. When the adults asked what was the matter, the child said, "The lights have come back. . . . They used to come to our old house all the time. . . . This is the first time the lights have come to our new house."

## 1990

One evening, at 10 P.M., a mother entertaining guests excused herself by saying, "I've got to go pick up my daughter. I'll be back shortly." She left her husband to regale their old friends with his tall tales. The woman should have been gone for thirty or forty minutes. By 11 P.M., the husband was concerned but hid his feelings. By 11:30, he called the home at which his daughter was visiting a friend, only to find his wife and daughter had left forty-five minutes earlier. By 12:30 A.M., fearing an accident had occurred, he was ready to call the police. Just then, the mother and daughter returned, oblivious to the fact that they had been gone for so long. "Did you see the moon?" the mother asked, almost in ecstasy. "Daddy, you should have seen the moon," the daughter said exuberantly. "It was unbelievable." One of the guests stepped into the backyard and looked at the sky. A gentle rain fell from a low cloud cover. It had been raining all day. There was no moon visible.

## 1996

One afternoon, an eight-year-old turned to a friend of his parents and said, "You won't believe what happened to me," the child spoke with innocent enthusiasm. "I came home yesterday and my nanny and my sister were in the kitchen and they were frozen. I tried to talk to them but they couldn't hear me. Then I turned and saw the aliens! The aliens asked me if I wanted to help them and I said okay. Did you know they only have four fingers? And the aliens have all different kinds of ships. Big ones to come here to Earth; little ones to fly around the sky; little ones to fly on their planet. Did you know the aliens have two planets right next to each other? They can see each other. And you can travel from one to the other. You just step into a tube and—poof!—you're there. And they showed me the future. It was scary. They asked me if I would help them and I said okay. Did you know the aliens have only four fingers? I think they were here last night but I'm not sure."

The youngster told a long, flowing, stream-of-consciousness tale filled with details about alien anatomy and technology that do not appear in the movies, television shows, cartoons, comic books, or computer games to which children are regularly exposed.

## 1997

An educated, accomplished businesswoman and her lover were vacationing in a cabin on a mountain lake. She had no particular interest in UFOs and was unfamiliar with the subject. Her companion had more than a passing interest in the subject. He had always wanted to see a UFO but had never witnessed one. About 2:30 in the morning, he sat up in bed. Out the window, over the lake, shone a brilliant pearl-colored orb. His partner was still sleeping as he got up from bed and went to the main room for a better view. As he stared in awe at the object hanging in the air above the lake, the woman awoke. "What is that!" she exclaimed. "I've never seen anything like it. It's so beautiful." Together, they looked at the light over the lake for about five minutes. They called out their observations to one another, each confirming what the other saw. Suddenly, the light vanished.

*Left: Daylight photo, the morning after the sighting of the UFO.*

*Next page: Artist's rendition of the sighting based on the description of the couple involved.*

A pair of high-quality binoculars lay on the table within inches of the man's hand, but he did not use them. His camcorder rested on the sofa a few feet away, yet he never thought to record the event. Frantic, he said to the woman, "We've got to remember this! Please, *you* must remember this. I may forget. I know it sounds crazy but I may not remember this."

"How could you forget something like this?" she asked, flabbergasted. "This is the most incredible thing I have ever seen."

They fell into a deep sleep shortly thereafter. The next morning, neither remembered the light on the lake.

None of the people involved in these stories want to be known. They do not want to appear on television. They are not writing books or magazine articles or making the rounds of the UFO lecture circuit. Every day, each of them goes about the business of daily life—earning a living, raising a family, getting an education. Each is an ordinary human being who has had an extraordinary experience.

Cynics often object to UFO reports by asking, "Why are UFOs only seen in rural areas in the dead of night by some uneducated Joe Six-Pack redneck?" The cynics prefer to put down the average person so they can dismiss their eyewitness testimony. However, increasingly, there are also well known, highly educated professionals who are coming forward to tell what they know, often at great personal risk to their reputations and careers.

John E. Mack, M.D., a Harvard psychiatrist and Pulitzer prize–winning biographer, came under an avalanche of criticism after the publication of his book, *Abduction: Human Encounters with Aliens*. Dr. Mack nearly lost his position at Harvard because he took the stories of abductees seriously.

Apollo astronaut Edgar Mitchell, Sc.D., has recently appeared before a congressional hearing and called for more government openness about UFOs. Dr. Mitchell believes that the secrecy surrounding the government's involvement in the UFO-extraterrestrial question has gotten completely out of hand and is a danger to our democracy.

Stanton Friedman—the man who resurrected the Roswell story from oblivion—is not an "apologist ufologist" but a scientist who asserts that we have all the proof we need to show that UFOs are real and that they are controlled by intelligent beings from outer space.

There are many other hard-working, serious investigators struggling to comprehend the UFO enigma, and untold numbers of sincere individuals who are grappling with the extraordinary experiences they have had.

But what do these experiences mean? What significance do they have?

Individually, these anomalous events probably have significance only for those who have experienced them. In general, each person interprets them in his or her own way. Some people are traumatized and suffer considerable pain; they must work hard to find equilibrium in life again after an encounter with aliens. Other men and women respond by embarking on a new path that brings greater meaning to their lives. However, none are quite the same afterwards.

Collectively, however, these experiences suggest that human beings all over the earth are experiencing something that cannot yet be explained by religion, science, or any other thought system yet devised. The UFO phenomenon is global in nature. Sightings occur on every continent virtually every day of the year. UFOs are reported by military pilots and preschool children, by members of Manhattan's ruling elite and by African bushmen, by religious zealots and confirmed atheists, by those with Ivy League Ph.D.s and those with high school GEDs.

The "modern era" of ufology began in 1947 with Kenneth Arnold's sighting in the Pacific Northwest and the Roswell incident in New Mexico and continues to this day. But the 1990s have been the Decade of the UFO. In the last six months of 1997, major publishers released more than two dozen books about UFOs. In the same year, the top grossing movie was *Men in Black*, which earned over $250 million. Of course, the highly acclaimed television series *The X-Files*, which has helped define the pop zeitgeist of the decade, has heightened interest and fascination in ufology and spawned a brand new genre of television show. Television commercials now regularly feature aliens and UFOs, a sure sign that the subject has infiltrated mass consciousness.

Despite the popularity of and interest in UFOs and extraterrestrial life, a strain of fear of the unknown and violence toward the different still runs deep in the human psyche and character structure. What is acceptable in the form of mass entertainment is still unacceptable as reality. For the vast majority of human beings on Earth, struggling to survive under harsh, grueling conditions, the question of extraterrestrial life is irrelevant. To philosophers, scientists and scholars, the topic is of increasing concern but the jury is still out. To those who have directly experienced the reality of life from elsewhere, terror is a common initial reaction. Many overcome this fear and move forward, but others do not.

We may not need to look to the heavens for proof of alien life. How can we doubt the existence of life on the billions of planets in the cosmos when we have evidence of "alien" life right here on Earth? On December 30, 1997, the *New York Times* ran a piece entitled, "Undersea Treasure, and Its Odd Guardians." The story described previously unknown alien ecosystems and the strange forms of life that abound in them, such as sea anemones that thrive on sulfur. According to the "newspaper of record," the deep sea is teeming with a profusion of life forms, the study of which will reshape our understanding of the evolution of life on Earth.

Humans first discovered alien life forms such as these in 1977 off the Galápagos Islands in the Pacific Ocean. Scientists have found that these life forms are not dependent on sunlight and that they thrive on chemicals, such as hydrogen sulfide, which would kill life forms that live on the Earth's surface. In short, these life forms are as alien to us as beings from another planet.

We know so little about Life on Earth. We are constantly surprised when our smug certainties and eternal verities are smashed and shattered by new knowledge. With such an explosion and riot of life here on our small planet, how can one imagine an empty universe devoid of Life's creativity and abundance?

And as we leave the Earth's surface, enter a new environment, and explore our solar system, how can we not understand that intelligent, curious beings on other worlds are not also likewise engaged? They, too, may be gazing at the cosmos right now, wondering if there is Life—perhaps even intelligent life—somewhere in the Milky Way.

Frightened cynics to the contrary, the case for UFO reality may not be as hard to make as one might think at first.

Innumerable suns exist;
innumerable earths revolve
about these suns. . . .
Living beings inhabit these
worlds.

—Giordano Bruno,
Sixteenth Century
Natural Philosopher

# CHAPTER ONE

# The Case for UFO Reality
The Case for UFO Reality
The Case for UFO Reality
The Case for UFO Reality
The Case for UFO Reality

E ons ago, the first creatures crawled out from the sea onto the land. These animals emerged from the only environment they had ever known to explore the unknown. They left their home to search. The Life Energy within them was reaching outward to make contact with a new environment, a new reality. We humans are part of this process which began billions of years ago.

Mere decades ago, human beings left the surface of Earth and began to explore space. The human animal left the only environment it has ever known, also to search. The Life Energy within continues to reach outward to make contact with a new environment, a new reality. A stunning, evolutionary leap is underway.

The men and women who are now venturing into space are part of an epochal advance equal to the development of Life on the surface of the planet. A living organism is taking its first tentative steps from the surface of the planet into the cosmic environment. Just as prehistoric animals left Earth's oceans to live on land, the human animal is leaving Earth itself and entering the cosmic ocean. Human exploration of space is not being done perfectly. There are dangers and errors. For example, human use of nuclear power in space exploration is potentially catastrophic and unnecessary; solar technology could be used instead. But the journey has begun and will continue.

What does this hold in store for us? Consider how vastly different our world is from that inhabited by those first creatures that ventured onto land. Who could have foreseen that *we* would develop from such beginnings? Yet, that is how different the cosmic world of the future will be from the reality we know today. Life in the future, after a billion years exploring the cosmos, will be as fantastically different from us as we are from our prehistoric ancestors.

Just as Life developed on Earth, and is now seeking to find Life beyond our planet, has Life developed elsewhere? Is such Life also searching and seeking? Might it already have found us? These are not new questions at all. They are almost as old as human thought itself. A few examples from the past should be sufficient to show that the question of Life beyond Earth has long fascinated humankind.

Ancient natural science posed questions that are asked today in fields as diverse as psychology and physics. The scientists of antiquity offered answers that still ring true concerning such mysteries as sensation, perception, self-perception, and consciousness. In the dawn of human reasoning and inquiry, people all over Earth looked up at the night sky and felt an intimate connection with the shimmering stars. These first cosmologists pondered the Universe and humanity's place in it.

Among them was the Greek astronomer Anaxagoras, who believed that there were worlds other than ours and that they were inhabited by living beings, including intelligent life forms who had developed civilizations. Anaxagoras believed there was no limit to the number of worlds with intelligent life on them.

Democritus, the great Greek astronomer and philosopher, also believed that there were an infinite number of worlds: some new; others in their prime; and still others in decline. Democritus felt that some of these planets harbored life but that others were devoid of the conditions necessary to support life.

Metrodorus, another Greek astronomer, wrote in the fourth century B.C.E., "To consider the Earth as the only populated world in infinite space is as absurd as to assert that in an entire field of millet, only one grain will grow."

Although many people now believe that Life exists only on Earth, nearly 2,400 years ago that notion was considered absurd. As political empires rose and fell, and Greece gave way to Rome, searching minds still sought to understand the cosmos. In the first century B.C.E., Lucretius, the author of the classic work *On the Nature of the Universe*, wrote, "The universe is infinitely wide. . . . Out beyond our world there are, elsewhere, other assemblages of matter making other worlds. Ours is not the only one."

In Asia, scholars were also studying the heavens and coming to similar conclusions. For example, in the thirteenth century C.E., Chinese philosopher Teng Mu wrote, "Heaven and earth are large, yet in the whole of space they are but as a small grain of rice. . . . How unreasonable it would be to suppose that, beside the heaven and earth which we can see, there are no other heavens and no other earths."

Science is not a steady progression forward. Much ancient knowledge was lost for millennia—such as the writings of the ancient Greeks—and other scientific contributions have not been able to survive attacks by religious or political forces threatened by scientific discoveries. For example, Aristarchus of Samos (310–230 B.C.E.), the Greek astronomer known as the "Ancient Copernicus," first proposed the heliocentric hypothesis of the solar system. He discovered that the sun is at the center of our solar system and that the planets revolve around the sun. However, this concept was lost for nearly 1,700 years. It was not until Polish astronomer Nicolaus Copernicus once again published the idea in 1543 that it reemerged in European thought. Copernicus, fearing the wrath of the Catholic Church, delayed publication of his scientific discovery for thirty years. His book did not appear until hours before his death.

In his book, *The Sleepwalkers*, Arthur Koestler refers to Copernicus as "the timid canon" for his delay in publishing his theories. However, Copernicus was correct to fear that his new astronomy might endanger his life. One of the most brilliant minds in human history—Italian philosopher Giordano Bruno—spoke out more freely about his revolutionary cosmology. In the sixteenth century, before the invention of the telescope, Bruno wrote, "Innumerable suns exist; innumerable earths revolve about these suns in a manner similar to the way the seven planets revolve around our sun. Living beings inhabit these worlds."

In 1592, Bruno was betrayed into the hands of the Inquisition. He was imprisoned and tortured for seven years. Bruno would not recant his cosmological views, which the Catholic Church considered heretical. On February 16, 1600, Bruno was burned to death by the heirs of Christ in the Square of Flowers in Rome. It is unlikely that the lesson of the murder of Bruno was lost on Galileo when, years later, he also came under attack. Galileo recanted his scientific findings when he was brought to trial by the Inquisition.

The Scientific Revolution that began in Europe with Giordano Bruno, Johannes Kepler, and Galileo Galilei, and which advanced with the discoveries of Sir Isaac Newton, René Descartes, and Albert Einstein, continues to this day. However, the mechanistic worldview these men helped to create is facing an unprecedented challenge, one that comes from beyond this planet—UFOs.

Accepting the idea that Life exists elsewhere in the Universe is not the same as believing that intelligent beings from other worlds are visiting our planet. But the fact that we are now exploring outer space suggests that this is possible elsewhere as well.

Are UFOs real? Is there hard evidence that UFOs exist? And, if there is evidence proving that UFOs are real, where do they come from? These are the basic questions that people the world over ask about unidentified flying objects.

Dr. J. Allen Hynek, one of the pioneers in ufology (who began his career as a debunker but became convinced of UFO reality by the mounting evidence), developed a terminology that is generally accepted in describing the main types of UFO experience. Hynek broke them down into six major categories:

1. nocturnal lights;
2. daylight discs;
3. radar/visual;
4. CE-I (Close Encounters of the First Kind), which describes a close observation of a UFO;
5. CE-II (Close Encounter of the Second Kind), which describes UFO incidents in which there is physical evidence present (*e.g.*, marks on the ground, physical effects on the

witnesses, electromagnetic interference with motors, etc.);

6. CE-III (Close Encounter of the Third Kind), a term, made famous by the Spielberg film of the same name, that indicates a UFO sighting in which alien beings are seen along with the spacecraft.

Since Dr. Hynek's death, the phenomenon of "alien abduction" has become more prominent and researchers have added a new classification—CE-IV (Close Encounters of the Fourth Kind)—to describe cases that involve such abduction.

## Documenting UFOs

The subjective reports of UFO witnesses require that the credibility of the witnesses be determined by asking questions such as: Can the person be believed? Does he or she have familiarity with known aircraft? Does the person know the terrain where the sighting was made very well? Credibility is an important criterion in evaluating a UFO report. But Dr. Hynek used another criterion when evaluating UFO reports as well—*strangeness*. The term "strangeness" is used to describe how greatly the sighted object differs from known aerial phenomena. UFO researchers look for sightings that have high levels of both credibility and strangeness. In this way, they are able to distinguish a true UFO from a rare natural visual phenomenon or unusual manmade object that is honestly misidentified.

There are also objective ways of documenting the UFO phenomenon. One reliable method is by radar. Throughout the modern era of ufology, radar has been used to verify visual sightings by witnesses. Experienced radar operators have provided convincing UFO evidence for the past five decades. In addition, UFOs have often left physical traces of their operations or landings. This evidence can be subjected to examination by well-known tools, such as the microscope or the Geiger counter. Also, photography can play a valuable role. Pictures can be taken of landing imprints from craft, changes in vegetation and soil where a UFO has been seen, and destruction of trees or plants from heat or radiation associated with a sighting. Chemical analysis can be applied to altered soil and vegetation as well.

The vast majority of subjective and objective UFO reports made over the past fifty years reveals that, to date, UFOs have four main shapes: disc-shaped, spherical, cylindrical, and triangular. Although some human aircraft have been manufactured in the shapes described, there has been little success in flying them. In addition, the performance capabilities possessed by the UFOs that witnesses have described, and that have been tracked by radar and captured on film and videotape, far surpass both known technology and even stories of the alleged exotic technology developed in secret government programs.

For example, the military has tracked UFOs traveling in the Earth's atmosphere at many thousands of miles an hour. They travel at speeds that would cause today's most advanced aircraft to actually melt from the heat generated. UFOs also have the ability accelerate while in flight, suddenly bursting to unbelievable speeds. Human aircraft cannot do this. The maneuverability of UFOs is spectacular and beyond anything present-day aircraft can achieve. UFOs can stop on a dime, make ninety-degree turns at extremely high speeds, ascend and descend thousands of feet in seconds, and zigzag while speeding through the atmosphere. And, finally, UFOs can hover in the earth's atmosphere. They can remain seemingly motionless, suspended in an eerie silence, for extended periods of time. Helicopters and other aircraft that can take off and land vertically produce loud roars that would sound deafening to any nearby witnesses.

In the United States, although UFOs are regularly sighted, they are only rarely reported in the mainstream print or electronic media. There is a kind of self-censorship in the American mass media concerning UFOs that does not exist in many other nations. On the occasions when a story does appear, people who report UFO sightings are all too often subjected to ridicule and contempt by the mainstream reporters. This treatment contributes to a reluctance on the part of the ordinary citizen to make a report of a sighting.

## Recent Reliable Sightings

To the average American, if he or she thinks about spaceships at all, it must seem that every few years one touches down in a remote part of the country and causes a minor stir that soon fades. The reality is quite different. A computerized data base— UFOCAT—maintained by the Center for UFO Studies in Chicago, Illinois, includes well over 50,000 reliable reports.

UFOs appear on an almost daily basis all around the globe. They appear over every continent and over every nation. For example, according to UFO researcher Budd Hopkins, a large UFO appeared over New York Harbor in the summer of 1997. There were multiple witnesses, and Hopkins is now investigating the matter. Yet the event received no coverage whatsoever in the media capital of the country.

On March 13, 1997, thousands of people witnessed a major sighting of a spectacular UFO (or of a number of UFOs) near Phoenix, Arizona, over the course of a number of days. At the time, stories appeared in the local press; then, months later, an article appeared in the June 18, 1997, issue of *USA Today*. Due to the popularity of the camcorder, dozens of people were able to record this amazing event, providing wonderful documentation of UFO sightings.

*USA Today* quoted one witness, a cement truck driver named Bill Geiner, as saying, "I'll never be the same. . . . Before this, if anybody'd told me they saw a UFO, I would've said, 'Yeah, and I believe in the tooth fairy.' Now I've got a whole new view. I may be just a dumb truck driver, but I've seen something that doesn't belong here."

Mr. Geiner experienced a mindshift.

These examples can be multiplied by the thousands. Here is a random sample of only some of the stories *CNI News*—an electronic magazine devoted to the UFO phenomenon—has run recently:

**UFOs Buzz Australia's Three Largest Cities**
**UK Press Told Not to Report Black Triangle UFOs**
**Numerous UFOs Reported Over Middle East**
**Brazil UFO Forum Highlights Global UFO Situation**
**Mother Ship Viewed at Close Range by Pilot**

**Top Mexican Newsman Details Aug 6 UFO Video Saga**
**Glowing UFO Seen Over South Korea**
**Dutch Abduction Reports Match U.S. Pattern**
**UFOs Active Over Scotland**
**More UFOs Reported Over Canada's Northwest Territory**
**Cuba Declassifies Reports of 1995 UFO Flap**
**Police Report UFO Over Lithuania**
**Police Videotape UFO Over Pretoria, South Africa**
**Hovering UFO Filmed by Soldier in Brasilia**
**Starship Vatican: Theologians Ponder a New Frontier**

These stories represent UFO news just for one year—and the headlines are barely the tip of the iceberg. UFOs are sighted over Australia, Korea, Africa, South America, the United Kingdom, Scandinavia, North America. The sightings are made by trained pilots, by police, by soldiers. The sightings are filmed and video-tape as well.

What do responsible, reliable individuals—people who are in positions to really know about UFOs—think about the subject? Their answers may be quite surprising.

General Nathan D. Twining, chairman of the Joint Chiefs of Staff from 1957 to 1960, wrote about his views on UFOs on September 23, 1947: "The phenomena reported is something real and not visionary or fictitious. . . . The reported operating characteristics. . . . lend belief to the possibility that some of the objects are controlled either manually, automatically, or remotely."

Admiral Roscoe Hillenkoetter, the first director of the CIA from 1947 to 1950, as reported in a 1986 article in *International UFO Reporter*, said that "Unknown objects are operating under intelligent control. . . . It is imperative that we know where UFOs come from and what their purpose is."

FBI Director J. Edgar Hoover, responding to a government request to study UFOs, wrote a note to his assistant, Clyde Tolson, on July 15, 1947, "I would do it, but before agreeing to do it, we must insist upon full access to discs recovered." Hoover's remarks are especially interesting, coming as they do immediately after the alleged Roswell crash and U.S. Air Force claims that no UFO crashed at Roswell, only a weather balloon.

The director of Project Magnet, the first Canadian government UFO investigation, Wilbert Smith, discovered in the 1950s that

"The [UFO] matter is the most highly classified subject in the United States, rating higher even than the H-bomb. Flying saucers exist. Their *modus operandi* is unknown but concentrated effort is being made by a small group headed by Dr. Vannevar Bush. The entire mater is considered by the United States authorities to be of tremendous significance."

General Douglas MacArthur was quoted in the *New York Times* on October 8, 1955, as saying, "Because of the developments of science, all the countries on earth will have to unite to survive and to make common front against attack by people from other planets. The politics of the future will be cosmic, or interplanetary." In 1962, in an address given at West Point, General MacArthur said, "You now face a new world—a world of change. The thrust into outer space of the satellite, spheres and missiles marked the beginning of another epoch in the long story of mankind. . . . We speak in strange terms of harnessing the cosmic energy. . . . of ultimate conflict between a united human race and the sinister forces of some other planetary galaxy."

A German rocket expert, Professor Hermann Oberth, considered to be one of the fathers of the space age, was quoted in 1954 as saying, "It is my thesis that flying saucers are real and that they are space ships from other solar systems. I think that they are possibly manned by intelligent observers who are members of a race that may have been investigating our earth for centuries."

In the excellent book, *Clear Intent* (republished in 1992 under the title *The UFO Cover-Up*) authors Lawrence Fawcett and Barry Greenwood quote from a chapter in a textbook in use at the U.S. Air Force Academy in the 1960s. The material was uncovered by NICAP, the National Investigations Committee on Aerial Phenomenon. In the textbook, *Introductory Space Science*, the last chapter is "Unidentified Flying Objects." The textbook summarized the major conclusions of the chapter for the reader. It is important to keep in mind that, at this time, the Air Force was telling the American public that UFOs were hoaxes, hallucinations, or honest misidentifications. What the Air Force was telling its own personnel was quite different: "From available information, the UFO phenomenon appears to have been global in nature for almost 50,000 years. . . . This leaves us with the unpleasant possibility of alien visitors to our planet or at least of alien controlled UFOs . . . what questionable data there are suggest the existence of at least three and maybe four different groups of aliens (possibly at different stages of development)."

In a 1978 letter to Grenada's ambassador to the United Nations, American astronaut Gordon Cooper wrote, "I believe that these extra-terrestrial vehicles and their crews are visiting this planet from other planets. . . . [W]e need to have a top level, coordinated program to scientifically collect and analyze data from all over the earth concerning any type of encounter, and to determine how best to interface with these visitors in a friendly fashion."

In 1996, Col. Cooper told Michael Lindemann of *CNI News* that he believes a variety of extraterrestrial are visiting Earth. There have long been rumors that the U.S. military recovered alien bodies in 1947 at the site of the alleged Roswell UFO crash. When asked about the recovery of alien bodies at Roswell, Cooper asserted, "I think there were better ones than Roswell. We've got some live ones. I know a guy who brought one in. An extraterrestrial who began to live among us."

Mercury 7 astronaut Donald "Deke" Slayton told the *National Enquirer* on October 23, 1979, that he had seen a UFO in 1951. Flying at an altitude of 10,000 feet above Minneapolis on a clear sunny day, he saw ". . . what looked like a saucer, a disc. About that time, I realized that it was suddenly going away from me— and there I was running at about 300 miles an hour." Slayton tracked the disc for awhile and then ". . . all of a sudden the damn thing took off. It pulled about a 45 degree climbing turn and accelerated and just flat disappeared." Slayton reported the sighting to military intelligence but never heard anything about it again.

Astronaut Edgar Mitchell, the lunar module pilot on Apollo 14, and later founder of the Institute of Noetic Sciences, spoke about UFOs in a lecture in New York City on December 4, 1991, sponsored by the Friends of the Institute of Noetic Sciences. He told those gathered, "I've changed my position in the last two or three years—the last two years to be precise—to suggest that the evidence is strong enough that we really need to have serious open discussion and release of information that it is quite clear the government and other governments do hold. . . ."

Apollo 15 astronaut Al Worden, in the PBS documentary, "The Other Side of the Moon," discussed the possibility of extraterrestrials visiting the earth in the distant past, as exemplified by the prophet Ezekiel's biblical UFO sighting: "A literal translation [of Ezekiel] describes very clearly a spacecraft with the ability to land vertically and take off vertically, and it was an object that looked very much like the Lunar Module we used on the moon. . . . I see some kind of intelligent being like us skipping around the universe from planet to planet as, let's say, the South Pacific Islanders do on the islands, where they skip from island to island. . . . I think we may be a combination of creatures that were living here on earth some time in the past, and having a visitation, if you will, by creatures from somewhere else in the universe. . . ."

Eugene Cernan, the commander on the Apollo 17 mission, has not seen a UFO himself. When asked about UFOs by a reporter, he was quoted in the *Los Angeles Times* on January 6, 1973, as saying, ". . . I've said publicly I thought they were somebody else; some other civilization."

Major-General Pavel Popovich, the president of the All-Union Ufology Association of the Commonwealth of Independent Sates, wrote in the *1992 MUFON International Symposium Proceedings*, that "The UFO sightings have become the constant component of human activity and require serious global study. In order to realize the scientific position of man on earth and in the universe, ufology, the scientific study of the UFO phenomenon, should take place in the midst of other sciences dealing with man and the world."

Soyuz 5 cosmonaut Yvegni Khrunov is reported to have said in the December 1980 issue of *Sputnik*, "As regards UFOs, their presence cannot be denied . . . their ability to change course by 90 degrees at great speed simply stagger the imagination."

In May 1981, while orbiting in the Soviet Salyut-6 space station, cosmonaut Vladimir Kovalyonok observed a UFO. He reported that it resembled a barbell. It seemed to be transparent and to contain another object inside, which he described as like a "body." The object seemed to be *pulsating, expanding and contracting.* Suddenly, there were two explosions, one half-second apart. "There were two clouds, like smoke, that formed a barbell. It came near me and I watched it . . . during a certain time, we and the craft were moving together." Major-General Kovalyonok made these comments in an videotaped interview near Moscow in 1993.

## Taking the Next Step

As this book is being written, and as this book is being read, human beings on every continent are interacting with nonhuman intelligences, extraterrestrials, aliens—call them what you will. Toward what purpose? To what end? To even begin to comprehend what is occurring, all preconceptions must be abandoned. The fearful thought, "That's not possible!" needs to be replaced by the exciting insight, *"Anything is possible!"* The barriers that separate human beings from full contact with reality need to fall. A radical shift in perception is required.

Fifty years into the so-called "modern era" of the UFO phenomenon, the mystery is as deep as ever. Although there have been at least 100,000 credible sightings and reports by unimpeachable witnesses worldwide, UFOs remain an enigma. Academic institutions, the military establishments of many nations, national and international governmental organizations, and a wide range of think tanks have all grappled with the accumulating evidence. Yet the nature of the phenomenon and its meaning remain unknown.

What is in the way? It could be that *we ourselves are in the way*. The character structure of the average human being—an individual's stereotypical way of acting and reacting—interferes with full perception of reality. Human beings have developed, over the millennia, a defensive "armor" that blocks the perception and expression of strong emotions and organ sensations. This armor is evident not only in rigid attitudes, but also, in rigid musculature. In many people, the mind and the body are in a constant state of defense against strong feelings that arise from outside and inside the self.

The UFO phenomenon produces intense emotions in those who experience it. The emotion may be awe or fear, wonder or terror, delight or dismay, love or hate. Anyone who has had a UFO sighting has experienced such intense emotions. However, when a rigid character structure blocks the emotions, the individual's experience becomes distorted in one way or another. Experience is then interpreted through the person's defenses. In this way, a true understanding of the UFO phenomenon can be blocked.

Rigid thinking is not only evident in individual human beings. It is apparent in entire societies as well. This is obvious when we study history and look back at past cultures. How often we smile at the obvious errors and ignorance of our ancestors. From the perspective of hundreds or thousands of years ahead in time, the limited and limiting belief systems of past eras seem so clear. Future generations will regard us in this manner.

Each era is marked by a particular mindset, a rigid set of limits on human thinking that inhibits further development. Thoughts that threaten the status quo are suppressed or stamped out. Often the thinkers of those thoughts are killed as well. This occurs today as much as in the historical past. A good analogy can be made between Giordano Bruno's fate in 1592, when he was delivered to the religious Inquisition, and the fate of the pioneering scientist Wilhelm Reich, who was under attack by the secular Inquisition in America in 1952. Bruno's books were banned by the Vatican, and he was burned at the stake; Reich's groundbreaking scientific and medical books were burned by the U.S. government and, in 1957, he died in a Federal penitentiary.

Our society is shaped by a mindset composed of mechanistic and mystical thinking. To the person with a mechanistic mindset or character structure, things function like a machine. The mechanistic scientist does not work with nature directly. For example, he kills cells and stains them before observing them under the microscope. Perfectionism is also an essential part of mechanistic thinking. Uncertainty, imprecision, and flux are anathema to the mechanistic mind. Nature, however, does not function like a machine. Nature has lawful functions but they are irregular, flexible, unstable. Mechanistic principles are useless when attempting to understand natural phenomena that are unpredictable, fluid, and nonmechanical—such as UFOs.

The mystical mind also reaches an impasse when it is confronted by the UFO phenomenon. In the mystical mindset or character structure, sensory impressions and organ sensations are blocked and then perceived as belonging to or coming from some realm "beyond." The mystical image of the world may be exact in every detail, but it is only as real as an image in a mirror. Reality itself remains elusive. The mystical mind does not apprehend the thing in itself. Therefore, the mystical mind believes the UFO to be a phenomenon from "beyond" or from a "spiritual" realm or another "nonmaterial" dimension of reality.

Neither the mechanist nor the mystic see a tangible connection between the emotional and the physical realms. Both are cut off from the direct perception of their own bioenergy and from bioenergetic contact with the natural realm. The UFO phenomenon involves natural energy processes that are inaccessible to both the mechanist and the mystic. Since mechanistic-mystical thinking is the dominant mindset of our era, the UFO phenomenon is not yet understood. It is a reality outside our framework of thinking.

It is extremely difficult to step outside the thought system of one's own time and see reality as it really is. Those who do so are frequently misunderstood by their contemporaries. The cosmology of Giordano Bruno and the science of Galileo threatened both the power structure and the character structure of the people of their day. The science of Wilhelm Reich likewise threatened the core beliefs of our era.

The UFO phenomenon, in a similar fashion, is outside the prevalent mindset and presents a powerful challenge to our basic assumptions concerning what reality is and who we are. The UFO phenomenon eludes the mechanistic-mystical mindset that now prevails. In a real sense, we cannot see what is right before our eyes. The mechanistic mind cannot comprehend something that does not function according to the laws of mechanistic physics; the mystical mind cannot grasp a physical reality that seems somehow beyond the physical, otherworldly, interdimensional, or "spiritual."

To understand the UFO phenomenon, a mindshift is required. Our thinking needs to change radically. The objective tools of mechanistic scientific thought and the subjective sensations of mystical thought are not adequate for the task. It is unlikely that some artificial fusion or amalgam of mechanistics and mysticism will do the job either.

In the field of ufology, there is very little solid ground. However, with almost 100 percent accuracy, we can be certain that the accepted wisdom of today will be seen in the future to be, at best, partially correct and, at worst, completely in error. In fact, most of the beliefs that we hold to so dearly and cherish with all our hearts are, quite simply, wrong. Although *we* are not aware of this yet, it will be painfully obvious to future generations. This cannot be helped. We are all limited by our social and personal circumstances. Therefore, it is crucial to keep an open mind and not be too quick to pass judgment on experiences that vary from our own. It is equally important to reserve judgment on extraordinary personal experiences that are uncommon in society at large.

Centuries ago, the poet Goethe wrote:

> What is the most difficult thing of all?
> That which seems the easiest:
> To see with your eyes
> What lies before your eyes.

Most people have lost the ability to trust in their own senses. They do not perceive reality through the unitary functioning of all five senses. Instead, they live in their heads observing the world through the intellect alone—as if the intellect does not depend entirely on sensory input. Consequently, when a phenomenon such as a UFO is observed, one that does not fit into their worldview, they actually cannot see it. Or, if the UFO sighting is so completely obvious, their minds work feverishly to "explain it away."

For example, an individual may be driving home and suddenly see an unknown object, three hundred feet in length, with glowing red, blue, and green lights, at treetop height off to the side of the road. Many motorists drive right on by, seeing nothing. However, this individual pulls off to the side of the road and watches the object for ten, twenty or even thirty minutes. It may be an awe-inspiring event, but it may also be deeply disturbing. The individual may experience an intense desire to deny the experience; it does not fit his or her picture of reality.

A UFO debunker may learn of the sighting and, without a visit to the site or any firsthand experience of the event, announce to the media that the three-hundred-foot-long, multicolored, solid object observed was really the planet Venus. The mystery is solved. Even the individual who saw the unexplained object may take the word of the debunker. As one of the Marx Brothers said, "Who are you going to believe—me or your own eyes?"

The fact that Venus is difficult to locate in the sky and appears to the naked eye as a tiny pinpoint of light is irrelevant. A way has been found to deny what one's own eyes saw. A way has been found to "explain away" disturbing sensory input and restore equilibrium to a mind troubled by a phenomenon that does not fit into its carefully constructed, but fragile, worldview. The fact that this is not a real solution does not matter.

Often it is easier to see our behavior in others. It is less painful to witness our shortcomings indirectly. John W. Mattingly, in his Foreword to *The Cancer Cure That Worked: Fifty Years of Suppression*, tells a fascinating story involving the indigenous people that the explorer Magellan met when he reached Tiera del Fuego in South America:

> *When Magellan's first expedition first landed at Tierra del Fuego, the Fuegans, who for centuries had been isolated with their canoe culture, were unable to see the ships anchored in the bay. The big ships were so far beyond their experience that despite their bulk, the horizon continued unbroken: The ships were invisible. This was learned on later expeditions to the area when the Fuegans described how, according to one account, the shaman had first brought to the villagers' attention that the strangers had arrived in something which although preposterous beyond belief, could actually be seen if one looked carefully. We ask how they could not see the ships . . . they were so obvious, so real . . . yet others would ask how we cannot see things just as obvious.*

How can *we*—the average person; the man in the street; the silent majority; the level-headed, sensible folk—not see what seems so obvious to millions of others around the globe? Are we like the residents of Tierra del Fuego, as limited and isolated by our fossil fuel-nuclear power culture as they were by their canoe culture? Do we have our own "shamans" who tell us of visitors from other worlds who arrive in preposterous vessels we call UFOs? Could we see these UFOs, too, if we looked carefully? And what would happen if we did?

Although the general public did not at first believe the stories of visitors from outer space, apparently the responsible leaders of the world did. The sightings were too real to ignore. The implications were too profound. The facts that the world leaders learned were so startling that they reacted by classifying them above top secret.

**PROJECT MINDSHIFT**

Photo by Wendelle Stevens.

Photo courtesy Photofest.

# CHAPTER TWO

# The Case for Government Secrecy
The Case for Government Secrecy
The Case for Government Secrecy
The Case for Government Secrecy
The Case for Government Secrecy

F rom September 1939 through August 1945, our planet was engulfed in World War II, the second global conflict of the century. The first forty-five years of the twentieth century were filled with slaughter—from the Boer War in South Africa, through the civil wars in Europe and Asia, to the Second World War. The peoples of Earth killed nearly one hundred million of their fellow human beings in a half century of carnage. The mass murder came to a temporary halt when the United States dropped atomic bombs on the Japanese cities of Hiroshima and Nagasaki in early August 1945, killing hundreds of thousands of civilians—men, women, and children.

The Age of the Atom Bomb had arrived, ushering in the Age of Anxiety.

A war-weary world struggled to emerge from the ruins of war, desperately hoping to enter an era of peace and cooperation. However, the military victory of the Allied forces—the U.S., Britain, France, and the U.S.S.R.—proved short-lived. The United States and the Soviet Union turned from allies into enemies, while the United States and Germany turned from enemies into allies. Political conflict in Europe, the Mediterranean, Asia, Africa, the Mideast, Central America, and South America erupted almost as soon as the military conflict ended. Small wars, civil wars, "brushfires," and "police actions" broke out all over Earth.

In the late 1940s, two great political systems clashed—Communism and Capitalism. They fought the battle for supremacy in nations large and small. People were the pawns in the deadly political games of brinkmanship that were played. Espionage, intrigue, subversion and assassination replaced the firebombing and nuclear annihilation of large cities. The hot war had been followed by the Cold War.

From high above our planet, all this suffering and destruction would have been invisible. From outer space, our tiny blue planet—a jewel in the solar system—would have seemed to spin tranquilly in its corner of infinity. Earth's loveliness and beauty would surely have attracted any cosmic voyagers passing this way. But how different Earth would have appeared to an extraterrestrial who stopped in for a visit. As Charlie Chaplin once observed, "In a long-shot, life is comedy; in close-up, it's tragedy."

Did any extraterrestrials come in for a closer look? There is evidence that they did indeed. They saw us as we really are . . . and we saw them. What did human beings actually see? What specific reports, made by highly credible witnesses, were so amazing that UFO information became *the most highly classified information in the United States?*

In the 1940s, during World War II, military pilots in the competing air forces reported similar unusual visual phenomena when they were in action in Europe and Asia. The pilots in the Allied forces called these visual phenomena "foo fighters." There is reason to believe that German and Japanese pilots saw these mysterious objects as well. The nature of the objects was not understood at all.

In mid-1944, at the beginning of the deployment of unmanned missiles, numerous sightings were made of glowing balls that seemed to be metallic. Many pilots reported that the "foo fighters" flew along with their planes, almost as if they were playing with them. The objects seemed to be under intelligent control and displayed no hostile activity.

A number of possible explanations for the "foo fighter" phenomenon were offered. Some military-oriented people believed that they were secret enemy craft. The Allied forces feared they belonged to the Axis powers and vice versa. Scientists suggested the pilots were actually reporting sightings of "St. Elmo's Fire"; that is, balls of static electricity. This phenomenon was often reported in the past, especially by sailors. (St. Elmo's fire plays a critical role in Herman Melville's great novel, *Moby Dick*.)

However, these explanations required that all the observations made by the great numbers of pilots and their crews be discounted entirely. Oddly, despite the great number of these reports, there is no available documented evidence that the military ever formally studied the phenomenon.

The development of the V-1 rocket by Nazi Germany, and later of the V-2 ballistic rocket, resulted in a rain of terror from the skies that caused tremendous destruction in London. The development of such rocket technology by the warlike inhabitants of Earth would surely attract the interest of intelligent beings from other worlds. It could be that the "foo fighters" reports were the first observations of UFOs in the modern era.

No sooner had the "foo fighter" phenomenon faded than a new, inexplicable aerial mystery took its place. Beginning in May 1946, "ghost rockets" were reported over Scandinavia. Witnesses described the objects in different ways: torpedo-shaped, cigar-shaped, or spindle-shaped. On June 9, 1946, over Helsinki, Finland, an extremely bright light streaked through the sky and left a trail that remained in view for about ten minutes. Nearly 250 reports came in at the time. The very next night, another strange light appeared in the sky—a light that seemingly *reversed its course.*

Unbeknownst to the public, government agencies were alarmed and began to investigate these sightings. In Europe, the first known UFO committee was established to look into the mystery. Military personnel were ordered to report sightings of the so-called "ghost rockets." On one July day alone, over 200 "ghost rockets" were reported. A great many of the sightings involved objects that flew low in the sky. Most of the objects made no sound or were barely audible.

The Swedish government formed a special committee to look into the unknown aerial phenomena. Soon after, the U.S. military became involved. The highest levels of the U.S. military were eager to conduct research and obtain solid factual information about the "ghost rockets." In August, stories about "ghost rockets" hit the mainstream press in Sweden and in the United

States. Hundreds of sightings were reported in the Stockholm area alone. Even the *New York Times*, the sedate and cautious voice of the ruling elite, ran stories on the phenomenon.

In a pattern that continues to this day, early reports in Sweden were followed by what can only be called either censorship of the press or self-censorship by the press. However, some stories did continue to appear in other Scandinavian countries for a while. Investigators were impressed by the great speed, the silence, the flight patterns, and the lights associated with the "ghost rockets."

Rumors began to spread that the mystery objects were secret Soviet weapons, developed from information captured from the Nazis. There was great fear that the Soviets had been able to develop advanced missile weaponry. This turned out not to be the case. The number of Scandinavian sightings waned, but reports came in from Hungary, Greece, Turkey, Spain, Portugal, and Morocco.

In 1978, a secret 1947 Defense Department document surfaced. The report stated that a Swedish Air Force pilot spotted a rocket, with no wings or rudder and leaving no exhaust trail, hugging the hilly terrain and maintaining a constant altitude, traveling at about 400 mph. Present-day cruise missiles are capable of such performance, but there was no technology available in 1946 that was capable of the "terrain-following" behavior reported by the Swedish military pilot.

In the summer of 1946, there were a few reports of "ghost rockets" in the United States. In the spring of 1947, metallic disks were reported on three occasions in Richmond, Virginia, by weather bureau employees. A sighting was also reported near Oklahoma City. Before the end of June 1947, unexplained aerial phenomena were observed on sixteen different occasions in the United States.

On Tuesday, June 24, 1947, the first American UFO "wave" began. (In UFO lingo, a *wave* is defined as existing when there are multiple nationwide or worldwide UFO sightings.) Businessman Kenneth Arnold was flying his single-engine private plane over southwest Washington State at an altitude of 9,200 feet, heading toward Mt. Rainier on a beautiful, clear, sunny afternoon. Mr. Arnold was involved at the time in a search for a missing military transport plane.

At one point, a bright flash of light reflected on his airplane. He turned and saw "a chain of nine peculiar looking aircraft." Arnold estimated the objects to be at about 9,500 feet elevation. It appeared to him that the objects were headed "in a definite direction of about 170 degrees." At first, Arnold assumed that he was seeing jet planes. Arnold noted that, every few seconds, two or three of the objects would change course and the sun would reflect off them. Arnold was puzzled by the appearance of the objects—he didn't see any tails.

Because the air was so clear, Arnold had no difficulty determining the size, shape, and speed of the objects. Arnold was intrigued by the seeming lack of a tail on the flying craft, and wanted a better look at the unusual objects. Arnold estimated that he was about twenty to twenty-five miles away from the objects. He knew that, to be seen so clearly at that distance, they must be quite large.

Arnold also had another well-known aircraft—a DC-4 airliner—in view at that time and was able to compare sizes. He judged the span of the strange aircraft to be the same as the distance between the farthest engines on either side of the DC-4 (about fifty-five feet). The mountaintops in the distance facilitated Arnold's estimate of the speed of the craft—1,700 mph. Such speeds were remarkable in 1947, when the world speed record for an aircraft was only 624 mph.

Arnold was not alone in sighting strange aircraft in the sky that day. Earlier, a half dozen discs were seen flying over the Cascade Range. One hundred miles to the east, near Richland, Washington, three unusual discs were sighted. And a man from the town of Mineral, Washington, who was positioned virtually directly under Arnold's Callair airplane, saw nine discs flying in formation at three o'clock in the afternoon. Arnold looked at his watch shortly after first noticing the flash of light and saw that it read 2:59 P.M.

When Arnold arrived in Pendleton, Oregon, word of his sighting had already preceded him. He was met by a curious crowd, among them a reporter for the *East Oregonian*, Bill Becquette. Arnold told Becquette that the craft all flew "like a saucer would if you skipped it across the water." From this statement grew the meaningless and misleading, but journalistically very popular term, *flying saucer*. The story was sent out over the Associated

Press newswire on June 25 in the late morning. Newspapers in the west carried the story in their evening editions. On June 26, Arnold's sighting hit the national press. Arnold was inundated with inquiries. Besieged, he returned home to Boise, Idaho.

On the Fourth of July 1947, multiple sightings were reported in the Portland, Oregon, area. Many papers carried UFO stories. The Army Air Force issued a statement denying that "flying saucers" were secret military technology. The military tried to explain away the sightings as meteors or ice crystals reflecting in the sun or huge, flat hailstones gliding through the air. At 9:15 P.M. on July 4, 1947, the flight crew of a United Airlines flight reported seeing five disc-shaped objects.

On July 6, 1947, the *New York Times* published a piece on flying saucers. In addition to running a list of possible explanations for UFOs (*e.g.,* weather formations, manmade objects) the newspaper of record suggested a rather daring possibility: "They may be visitants from another planet launched from spaceships anchored above the stratosphere."

For Kenneth Arnold and the American public, the era of the flying saucer had begun. Simultaneously, the era of the "debunker" had begun as well. From the very beginning, "experts" appeared on the scene determined to explain away the observations made by people they had never met, concerning events they had never seen that had occurred in places they had never been. These individuals invariably described themselves as skeptics. No description could be more in error. A skeptic is an open-minded seeker of the truth, no matter what that truth may be. Debunkers, however, begin their research with conclusions already drawn. They do not seek to understand a situation or to uncover the truth of an event, but only to explain away what does not fit into a tenaciously held, but easily toppled, mindset.

Debunkers almost always have two things in common. First, their explanations of UFO sightings and UFO-related incidents are far more preposterous than the notion that intelligent beings are exploring our solar system. And, second, their fierce determination to explain away the unknown and disparage those who report UFO sightings masks a deep fear and insecurity. Only an individual whose psychological and intellectual underpinnings are unstable would devote his or her work life to debunking the perceptions of others instead of performing creative work of his own.

To a person, all of the major UFO investigators state without hesitation that no one knows the full truth about UFOs. The only ones who claim complete, certain, full knowledge are the so-called debunkers.

The attempts to explain away what Arnold saw began immediately. But what few people were aware of was that Arnold's sighting was not really the first one of the modern era at all. It was simply the most widely reported sighting. Among the earlier sightings were:

- April 1947: A weather bureau employee in Richmond, Virginia, Walter A. Miczewski, and fellow employees noticed a silver disc cross the sky, heading from the east to the west.

- May 1947: A silvery object fell from the sky over Washington State and apparently disintegrated; observers in Richmond, Virginia, reported a cigar-shaped object head northwest across the evening sky; a silvery object was reported in Manitous, Colorado.

- June 1947: Coral Lorenzen saw a spherical light at Douglas, Arizona, rise and fly into the sky; many UFOs were reported flying together over Weiser, Idaho; a triangular formation of ten UFOs was reported in the afternoon over Bakersfield, California; eight flying discs were reported just before noon near Spokane, Washington; a sighting at dusk was reported in Yukon, Oklahoma.

On the very day of the legendary Arnold sighting, Fred M. Johnson saw six discs flying in the sky about twelve miles from Mt. Rainier in the Cascade mountains. Johnson may have witnessed the same craft as Arnold. Later, on June 24, near Seattle, Washington, between 9 and 11 P.M., dozens of witnesses saw blue and purple lights in the sky in formation.

In 1968, UFO researcher Ted Bloecher reported that there were 832 sightings between June 15 and July 15, 1947. Today, it is estimated that the total number of sightings from that period—including reports from outside the United States—is close to 1,500. The official U.S. Air Force files for the period, as reflected in Project Blue Book, list only fifty UFO reports from the same period.

A 1947 Air Force study, the first known official UFO study, came to some very significant conclusions. Among them were:

- the flying saucer phenomenon is real;
- the President knows about this situation; and
- the reported objects
  1. appear metallic and possess at least a metallic skin;
  2. when the objects emit a trail, it appears as a blue-brown haze;
  3. all observations to date indicate circular or elliptical objects, with a flat bottom and domed top;
  4. the objects are estimated to be the size of a C-54 or Constellation (wingspan 120 feet; length, 95 feet).

From the very beginning, the U.S. military and the U.S. government took the UFO matter seriously. The behind-the-scenes reality, however, did not match the public posturing. The public was always led to believe that there was nothing to the flying saucers.

Somehow, this policy of deception escaped the notice of the public relations officer at an Army Air Force base in New Mexico. Inadvertently—or perhaps *advertently*—this public relations officer issued an official press release on July 8, 1947, that rocked the official boat all the way from a sleepy New Mexico desert town to the innermost echelons of power in Washington, D.C.

## The Roswell Incident

Lieutenant Walter Haut, the public relations officer for the 509th Bomber Group at Roswell Field, the world's only atomic bomb group, startled the world and the powers that be by announcing that a crashed disc had been retrieved by the Army Air Force at a site outside of Roswell, New Mexico. The story was denied, the military retracted the claim, a successful cover story about a crashed weather balloon was accepted by the press, and the story sank beneath the arid desert sands. The "Roswell Incident" remained forgotten until 1978 when UFO researcher Stanton Friedman succeeded in getting the first eyewitnesses—Jesse Marcel, Sr., Walter Haut, Glenn Dennis, and others—to speak about what happened to them and what they saw thirty-one years before.

Today, *Roswell* has become a buzzword connoting UFOs, extraterrestrials, crashed discs, recovered alien bodies, back-engineered alien technology, and government cover-up of the truth. But for three decades, what many believe to be the story of the century was successfully buried. There are many excellent books on Roswell available—notably *The Roswell Incident* by Charles Berlitz and William Moore, *Crash at Corona* by Stanton Friedman and Don Berliner, and *Top Secret/Majic* by Stanton Friedman—that tell the full story of the UFO crash and convey its significance. In fact, with the fiftieth anniversary of the Roswell Incident in 1997, a whole Roswell cottage industry of television shows, dramatic films, documentaries, magazines, and books has sprung up. Therefore, the reader is urged to turn to the literature for the details concerning Roswell.

However, it is worth considering some recent information about Roswell reported by Jim Marrs in his 1997 book, *Alien Agenda*. In 1947, the Army Air Force claimed that the wreckage found at Roswell was a secret weather balloon, not a crashed spaceship. In 1994, the official explanation was that the wreckage at Roswell was not a just any weather balloon. It was a *really, really secret* Navy Skyhook weather balloon. The official U.S. Air Force explanations reached a farcical nadir in 1997 when, on the fiftieth anniversary of the Roswell incident, military officials claimed that the alleged Roswell alien bodies were actually "crash test dummies" from a top secret project. What makes the government's assertion patently absurd is that the Roswell bodies were claimed to have been seen in 1947. Crash test dummies were not used in secret projects until 1953—six years later.

According to Marrs, there is no doubt that *something* crashed at Roswell in July 1947. Since the first eyewitnesses spoke, the list has grown. According to Marrs, among those who have told what they know about Roswell are as follows:

William Woody
Mother Superior Mary
  Bernadette
Sister Capistrano
Cpl. E.L. Pyles
James Ragsdale
Trudy Truelove
James Ridgway
C. Curry Holden
Dr. C. Bertrand Schultz
Maj. Jesse Marcel
Dr. Jesse Marcel, Jr.
Col. William Blanchard
Maj. Edwin Easley
Sgt. Thomas C. Gonzalez
Steve MacKenzie
Lt. Colonel Albert L. Duran
W.O. Robert Thomas
M. Sgt. Bill Rickett
Sgt. Melvin E. Brown

Frank Kaufmann
W.O. "Pappy" Henderson
Sarah Holcomb
Helen Wachter
John Kromschroeder
Maj. Ellis Boldra
Floyd and Loretta Proctor
William Proctor
William W. "Mac" Brazel
Bill Brazel
Sallye Tadolini
Frank Joyce
Glenn Dennis
E.M. Hall
George Bush (not the
  president)
Sheriff George A. Wilcox
Barbara Dugger
Frankie Rowe
Brig. Gen. Arthur E. Exon

Marrs correctly points out that if, for argument's sake, half of the above witnesses are thrown out, the remaining testimony is enough to give a person with an open mind reason to consider seriously what they have to say.

It is often asked how it would be possible to keep such a secret. It is hard for many people to believe that an event of such magnitude could be hidden from the public. It may be easier than it seems at first. For example, Roswell is not too far from Los Alamos, the center of the American effort to build the atomic bomb during World War II. Los Alamos grew to a city of thousands of people during the war effort. Yet, the work on the atom bomb remained a complete secret. In fact, the very existence of Los Alamos itself remained a complete secret. The Manhattan Project—the code name for the effort to build the atomic bomb—was also a secret, even to the scientists who were working on it.

If these two examples are not convincing, consider this. Stanton Friedman, the man who began the unraveling of the Roswell Incident, was once a nuclear physicist working on the U.S. space program. Friedman freely admits that he himself still keeps secrets concerning classified information pertaining to his former work for the U.S. government. If the man who broke the Roswell story keeps some information secret, is it so surprising that the eyewitnesses kept silent under direct order or threat of reprisal if they told what they knew?

It is easier to keep secrets than people are willing to admit, for almost everyone has his or her own secrets. But societies and individuals would do well to heed the words, "You're only as sick as your secrets."

## The Politics of Secrecy

Should it seem that unwanted information is about to get out, bureaucracies have their ways of preventing such eventualities. In response to pressure from his constituents, the late U.S. Congressman Steven Schiff (R–N.M.) requested the Government Accounting Office (GAO) to search for documents concerning the Roswell incident. In July 1995, the GAO reported its findings: "The Roswell Army Air Field administrative records from March 1945–December 1949 were destroyed, as were ongoing messages from October 1946–December 1949." Schiff's office said the congressman was surprised that the documents were destroyed, apparently without authorization, or at least without any records of authorization.

The destruction or alteration of public records may have surprised the congressman, but it is, sadly, a familiar experience to those who seek to force government agencies to yield information that belongs not to the bureaucracies, but to the people.

For Marrs, Roswell marked a turning point in military attitudes toward the UFO phenomenon. Before Roswell, he asserts, the military was "intensely interested in UFOs and open to the idea that they represented extraterrestrial visitation." After Roswell, an American "Iron Curtain" of Secrecy descended and hid the available UFO information from the public.

Two months after Roswell, the National Security Act of 1947 became law. The National Security Council was formed. In a purely Orwellian turn of the phrase, the Department of War became the Department of Defense. The Army Air Force, which had bungled the Roswell Incident at first, became the independent U.S. Air Force. At the CIA, Alan Dulles and company were secreting Nazi scientists and psychiatrists into the United States, and hiring thousands of former Nazi espionage agents to work for the CIA. The national security state was spreading its dark cloak over the sleeping American democracy.

The cult of secrecy was ensconced in power. Nazism had been defeated militarily, but international fascism was spreading its tentacles into American corporate and governmental life. However, the American public was otherwise engaged. It was busy with baseball; or building a new career in the wide-open fields of advertising, television, marketing, and public relations; or buying that first new home with the white picket fence that is each American's birthright; or getting behind the wheel of that new, bright red, gas-guzzling Chevrolet and heading off to see the USA; or starting a family and ushering in the baby boom.

It was the American people who turned their lives over to others, who gave those who ruled in their stead the power to do so in secrecy.

Secrecy can sometimes result from the pathological character structures of people in power. And some of the UFO secrecy surely can be traced to such factors. At other times, secrecy may be necessary in the face of external events. In the 1940s, irrational international politicians now had atomic weapons. The world's leaders were only reflections of those they ruled. The people of Earth had created the conditions that put humankind in danger of annihilation.

Within the mindset of the time, secrecy made sense. Within the social situation of the time, espionage and military preparedness, bluster and brinkmanship all made sense. At that period of history, with the conflict between the state capitalism of the Soviet Union and the private capitalism of the United States threatening to break out into nuclear war, all kinds of unspeakable acts made sense—but only when one accepted the assumptions of that mindset.

If one enters into the mindset of a schizophrenic, for example, and accepts the schizophrenic's basic assumptions, the whole distorted view of reality makes sense. But, of course, only within that distorted "worldview"; if one steps outside of the situation, one is able to see it as it really is.

However, it was not only irrational international politics that confronted the world's leaders. Human character structure (or "human nature," as most leaders would have then considered it) was at the root of the dilemma. Human fear of the unknown and hatred of the different could be catastrophic in the face of life from elsewhere. The emotional response of Earthlings to intelligent life from outer space posed a grave potential problem.

In America, people with black skin could not even drink from the same water fountain as people with white skin. In the Middle East, Semitic peoples who prayed to the god they called Allah and Semitic peoples who prayed to the god they called Jehovah wanted to murder one another. In China, centuries of rigidity and oppression were yielding to a rage that would bring death to tens of millions. In the Soviet Union, a cruel form of oppression was tightening its grip and extending its reach into Europe.

The world's political leaders lacked any insight into the depths of human character structure. Politicians in all lands certainly knew how to manipulate human emotion and twist it to exploit people for their own power-political ends. But they knew it was a dangerous game. In the face of their ignorance about people, and in light of people's destructive and self-destructive behavior, they came to the conclusion that they could not trust the citizens of their nations with the truth about UFOs as it was then known. The leaders feared panic and chaos would follow the revelation that Earth was being visited by superior beings from another world.

The leaders of the day, at first, were justified in keeping the lid on the UFO phenomenon. Or, if they were not justified, their behavior is at least understandable. The U.S. military did not know if the mysterious aerial phenomena were foreign military technology. It had to be determined with certainty that UFOs were not advanced Soviet weaponry.

If UFOs were not human-made, one problem was solved but a bigger one loomed. What if the UFOs were from other planets? The military and political leaders didn't know what UFOs were, where they came from, or why they were here. Government leaders did not know if they were hostile, benign, or indifferent. In a sense, despite all their might, the leaders were powerless.

In 1947, what could the American public have been told by the military? That unknown objects were entering their airspace with impunity, flying over their most secret atomic research centers and military bases, and that the government had no idea where they had come from or what they wanted? That was certainly not a possibility.

As a result, the leaders took the only course they thought was available to them at the time. They kept silent when possible. They hid the truth. They covered up whatever facts somehow became public about UFOs. Faced with their own ignorance and fear, the powerless people in power used lies, ridicule, threats, and defamation to buy themselves some time.

At the end of 1947, the world scene was poisoned by suspicion and fraught with danger. Strange craft were being seen in the skies. The objects flew at speeds and maneuvered in ways that were beyond the capacity of any known technology. Some believed the mystery craft were foreign technology that posed a threat to their nation. Others suggested that they were extraterrestrial spacecraft whose purpose remained hidden.

It was in this atmosphere that the first of a series of secret UFO projects was begun by the United States government.

True story behind the strangest phenomena in history!

107

GOLD MEDAL BOOK

# THE FLYING SAUCERS ARE REAL

DONALD KEYHOE

Fawcett Publications, Inc., 1950.

The astounding FACTS behind the greatest mystery of our age — by the recognized authority

★ 207

SMITH Family Book Store

USED

# FLYING SAUCERS FROM OUTER SPACE

## Major Donald E. Keyhoe
U.S. MARINE CORPS. (Ret.)

Complete and unabridged

Permabooks (Doubleday) edition, 1954. Henry Holt and Company, Inc., 1953.

PROJECT MINDSHIFT

# CHAPTER THREE

# Government Investigations
# Government Investigations
# Government Investigations
# Government Investigations
# Government Investigations
# —The Secrecy Deepens

I n 1948, the U.S. military was making major technological advances in the areas of atomic weaponry, rocket science, and radar. This work was centered in southeast New Mexico, a beautiful part of Planet Earth. To some, however, it seems a sad and sorrowful beauty. Something in the land is dying in this part of America. It was here that human beings came to develop and deploy weapons of unparalleled destruction.

At Alamogordo, New Mexico, the world's first atomic explosion occurred. A short drive south of Alamogordo, on Highway 70, lay the White Sands Testing Range, where top secret American rocket technology was being developed and tested. The 509th Bomber Group—the only atomic military unit on Earth at the time—was stationed at the Roswell Army Air Force Base just over one hundred miles east of Alamogordo. In addition, vastly improved radar systems were being devised in the military bases in this area. A few hours to the north, in Los Alamos, beat the heart of the American atomic program.

If intelligent beings from other worlds were monitoring Earth, the explosion of an atomic bomb would certainly have drawn attention. In addition, the successful use of rockets in war could scarcely have escaped notice. And more effective radar systems would require an increased defensive response from spacecraft entering Earth's atmosphere. All three technological developments—atomic weapons, rockets, and advanced radar—could be investigated in the southeastern section of New Mexico.

## Project Sign

All of the above coincided with the advent of the National Security State in the United States. Project Sign was the first UFO investigation in America. It began operations in January 1948. Project Sign had a high national priority rating, but a low security classification. From the start, there was division among the ranks, with some believing UFOs were Soviet technology, others that they were extraterrestrial.

Dr. J. Allen Hynek joined Project Sign in early 1948. Hynek was not impressed by the 200-plus reports that Project Sign first studied. At first, Hynek considered the whole topic of UFOs to be nonsense—something to laugh at, a subject of interest only to the gullible and naive. He joined Project Sign hoping that the application of the scientific method would prove the mysterious objects to be nothing more than misidentifications of natural phenomena or manmade objects, or the work of hoaxers. To Hynek, UFOs were figments of the imagination.

Hynek's views were shared by others involved in Project Sign. Then, on January 7, 1948, the investigation of a UFO sighting led to the death of a pilot, Kentucky Air National Guard Captain Thomas F. Mantell, Jr.

At about 1:30 P.M., reports of a large white object with a red light began to come in to the Kentucky State Police from Maysville, Irvington, Madisonville, and Owensboro, Kentucky. The object was visible to the control tower at Godman Air Force Base. Captain Mantell was flying with three other planes; they were asked to investigate. One pilot was excused but Mantell and two others began the investigation. Mantell radioed in, "The object is directly ahead of and above me now. . . . It appears to be a metallic object or possibly reflection of sun from a metallic object and it is of tremendous size. . . . I'm trying to close in for a better look." Mantell and the other pilots climbed higher into the sky toward the mysterious circular object.

Mantell's companions turned back at 20,000 feet but Mantell continued to climb. The WWII F-51s were not equipped with oxygen. At 30,000 feet, his plane leveled off and then began a spiraling dive. Mantell's body was removed from the crash site. His stopped watch read 3:18 P.M. Reports of the UFO continued for some time after Mantell's death.

On January 9, 1947, the *New York Times* reported Mantell's death. The Air Force claimed Mantell died chasing Planet Venus. In his book, *Watch the Skies!*, Curtis Peebles concludes that Mantell passed out at 25,000 feet from lack of oxygen but he does not present the data he used to reach that conclusion. Commenting on the Air Force's belief that Mantell was chasing Venus, Peebles remarks, "The one problem was that Venus was at only half its maximum brightness, making it hard to see against the bright sky." What makes Peebles observation particularly interesting is that he believes that UFOs are misinterpretations of conventional objects and that the subject is nothing more than a myth.

However, according to Jim Marrs, the Godman Air Force base commander told reporters he had observed the UFO for nearly an hour with binoculars. It is unlikely that he was observing Venus for sixty minutes. Marrs also reports that a man named Richard T. Miller was in the operations room of Scott Air Force Base in Belleville, Illinois, during the Mantell incident. Miller claims he heard Mantell's last radio transmission, "My God, I see people in this thing!"

As with almost every UFO incident, there were far more questions than there were answers and there was great controversy about what actually happened to Captain Mantell. Project Sign eventually said that Mantell had not been chasing Venus, which was at the time a tiny pinpoint of light far below Mantell's craft, not above it. Instead, the official version was that Mantell was chasing a secret Navy Skyhook balloon. (The regularity with which trained, experienced military personnel mistake UFOs for various kinds of balloons strains credulity.) In the days following the death of Captain Mantell, a similar UFO or perhaps even the same one, was reported over North Carolina.

On July 24, 1948, Captain Clarence Chiles and his First Officer reported a UFO sighting made while they were piloting an Eastern Airlines DC-3 to Atlanta, Georgia. The plane was flying at about 5,000 feet. A bright moon was shining, and there was broken cloud cover above the airplane. At about 2:45 A.M., the pilots saw a wingless, cigar-shaped craft on the starboard side. As it approached them, they had assumed it was a jet plane. Chiles said that the craft seemed to be about three times the size of a B-29 Superfortress. There appeared to be two rows of windows, with a bright light glowing within. There was a blue glow underneath the ship. The craft was in view for five to ten seconds. Both aircraft took evasive action. Chiles thought the UFO was piloted. One passenger also saw the UFO.

Project Sign officers took the sighting seriously and, after investigating it, concluded the craft was extraterrestrial. The project officers completed a report, called "Estimate of the Situation," in September 1948, stamped it "Top Secret," and sent it to higher-ups for review. No one suspected that this report would become one of the most controversial documents in a volatile field.

The report acknowledged that Arnold's was not really the first UFO sighting. They provided other reports from such reliable witnesses as pilots and scientists. Their conclusion was that UFOs were real and that they were extraterrestrial. General Hoyt S. Vandenburg, the U.S. Air Force Chief of Staff, received "Estimate of the Situation" in October 1948. He rejected the document's conclusions in only a matter of days. All copies were ordered burned. But the book-burning backfired. Word got out, and interest in the document increased dramatically. The destruction of the document was seen as evidence of a government cover-up.

In February 1949, the Project Sign staff completed their final report. It was classified "Secret." The report noted that, "All evidence so far presented on the possible existence of space ships from another planet" has been largely a matter of conjecture. However, in its analysis of the Chiles sighting of the cigar-shaped UFO, the Project Sign staff concluded that ". . . it is probable that the method of propulsion of this type of vehicle is far in advance of presently known engines."

The Project Sign report included an analysis by Professor George E. Valley of the Massachusetts Institute of Technology and the Air Force Scientific Advisory Board. Professor Valley wrote, "If there is an extraterrestrial civilization. . . . Such a civilization might observe that on Earth we now have atomic bombs and we are fast developing rockets. In view of the past history of mankind, they should be alarmed. We should expect at this time, above all, to behold such visitations."

A second scientific point of view was provided by Dr. James E. Lipp of the Rand Corporation. He wrote, "The first objects were sighted in the Spring of 1947, after a total of 5 atomic bomb explosions, *i.e.*, Alamogordo, Hiroshima, Nagasaki, Crossroads A and Crossroads B. . . . " Lipp concluded, "the spacemen are feeling out our defenses without wanting to be belligerent." In addition, Lipp believed that the majority of the stars had conditions suitable for the development of life. Dr. Lipp was not in tune with his scientific peers. To these minds, the idea of the existence of advanced alien civilizations was tolerable, but the thought that such civilizations may be visiting earth was difficult to accept.

One part of the staff of Project Sign thought they had done their job well. They had successfully "explained away" all but seven of the cases assigned to them. In reality, they hadn't actually explained anything at all, but that seemed beside the point. The staff members who supported the extraterrestrial hypothesis were discouraged by the course of events.

The news media eventually discovered the existence of Project Sign (even a broken clock is right twice a day). They inaccurately referred to it as Project Saucer. In February 1949, the covert project was renamed Project Grudge. According to Dr. Hynek, it was at this time that the Pentagon began to treat the subject of UFOs with "subtle ridicule."

## Project Grudge

In secret, Project Sign had concluded that UFOs were extraterrestrial. But the public would not learn that for decades. In contrast, Project Grudge made every effort to promote its conclusions—that all UFOs could be explained away—through a carefully constructed public relations campaign. Hynek noted the change in attitude. He even later wrote that the statements that Project Grudge made about particular UFO cases, through its public relations campaign, bore little resemblance to the facts.

As the UFO data became more significant, an intense effort was also underway to dampen and extinguish all interest in the phenomenon. This confused many who took the matter seriously. Captain Edward J. Ruppelt, who later headed his own UFO investigation, referred to the period in which Project Grudge operated as the "Dark Ages." In general, the reports of Project Grudge were ambiguous at best. At worst, they were transparent efforts to explain away the phenomenon. The staff's disregard for the facts—especially the most obvious fact that they had not solved a good number of cases—was evident to anyone who took the time to read the reports (which few did).

According to Ruppelt, military officials at the Air Technical Intelligence Center regularly ridiculed the whole subject of UFOs—unless you were alone with them. In a startling turnabout, these mocking debunkers frequently became quite serious about the subject of UFOs and, on occasion, even argued with others who tried to laugh the whole thing off.

For many reasons, such as underfunding, lack of staff, and its own ambivalent attitude, Project Grudge did not get the job done. The American public was still avidly interested in flying saucers, as they were still called then, and the effort underwent a major reorganization. On October 27, 1951, "the new, improved" Project Grudge officially began, with Captain Ruppelt in charge.

Under Ruppelt's leadership, the situation improved dramatically. Better reports were made. In addition, Air Force pilots and air traffic controllers received orders on the proper procedures to follow after a sighting. These procedures were known officially as CIRVIS—Communication Instructions for Reporting Vital Intelligence Sightings.

Unfortunately, existing joint Army-Navy-Air Force reporting procedures contradicted the Air Force guidelines. The new Air Force policy even contradicted one of its own regulations. Bureaucracies at odds with one another—even a bureaucracy at odds with itself—is the nature of the beast, understandable to any citizen who has had even the most benign contact with government agencies. However, there were serious consequences to noncompliance with these military policies.

JANAP (Joint Army-Navy-Air Force Publication) 146 ordered pilots to make reports of UFOs, but only in secret. Failure to comply would put the pilot in violation of the espionage laws and the Communications Act of 1934. And Air Force Regulation 200-2 kept all UFO reports under the purview of three special Air Force intelligence groups.

In February 1954, this secrecy was extended to include commercial airline pilots. The Air Force instructed all commercial pilots to send UFO reports to the intelligence department of the Military Air Transport Service (MATS) in Washington, D.C. Pilots were further "asked" by the military to refrain from speaking to the press about what they witnessed and not to discuss their sightings publicly in any way. Hundreds of pilots protested this decision and signed a petition to that effect. However, powerful people behind the scenes were determined to control the flow of UFO information; nothing came of the pilots' protest.

Major Donald Keyhoe—a major player in the early days of ufology and the author of such books as *Flying Saucers from Outer Space*, *Flying Saucers: Top Secret*, *The Flying Saucers Are Real*, and *Aliens from Space*—spoke out forcefully against Air Force secrecy and suppression of information. According to Keyhoe, JANAP 146 and AFR 200-2 had the immediate result of keeping hundreds of new UFO sightings secret. While military leadership insisted flying saucers did not exist, the military machine seized control of all information and blocked its release as effectively as it could.

This secrecy is not a thing of the past; it continues to this day. An Air Force serviceman, now retired, who flew on thermonuclear bombers in the 1960s and 1970s, awaiting orders "to take out some Russkie city," as he put it, said that "we saw UFOs all the time. But we can't talk about it."

The keepers of the secrets faced a serious challenge when, on August 25, 1951, strange bluish lights were seen over Lubbock, Texas. The lights lasted for several consecutive days and could be seen all over west Texas. A chemist, a physicist, a geologist, and a petroleum engineer—all from Texas Technical College—witnessed the lights on successive nights. These academicians concluded that the lights were under intelligent control.

The so-called "Lubbock Lights" were photographed by Carl Hart, Jr., a college freshman. He got five images of a V-shaped formation which the Air Force tried to dismiss as the reflection of light from streetlamps on the rear ends of birds. The residents of Lubbock apparently felt this explanation came from the rear ends of the Air Force officials. They knew what they had seen with their own eyes.

Ruppelt investigated the Lubbock Lights incident for Project Grudge; the bird story didn't fly with him, either. He ruled out any role for streetlamp light since the objects were also seen in remote areas far from town. Ruppelt knew that the Lubbock Lights had been seen on radar from as far away as Albuquerque, New Mexico. Many government scientists believed the Lubbock Lights were interplanetary spacecraft. (Five years later, based on "secret" information from an "unidentified source," Ruppelt concluded the lights were merely a natural phenomenon.)

In the history of government UFO investigations, lifelong residents of areas of the country—people with no previous interest in UFOs, acting either as individuals or in groups—are deemed to be suddenly so confused or seized with inexplicable hallucinations that they identify ordinary natural occurrences as interplanetary spaceships. Or trained military and commercial pilots, who have thousands of hours of flying time—people in whom the lives of millions are entrusted—supposedly, suddenly and inexplicably, think that the moon or Venus or another ordinary celestial sight is a spaceship from another planet. These common government explanations for UFO sightings are far less believable than the idea that ordinary people have witnessed something extraordinary.

Ruppelt had tried to be fair-minded with Project Grudge. He dismissed some staff members who had proved themselves not open to considering new facts, both "skeptics" and "believers." In January 1952, Ruppelt and a Grudge scientist met with employees of General Mills who claimed to have seen a UFO. These individuals were part of the company's Aeronautical Division and were knowledgeable in aerodynamics, astronomy, and meteorology. In addition, they had been involved in *launching and tracking every Skyhook balloon the military had launched prior to 1952.* Their testimony to Ruppelt shattered the foundation of one of the military's main explanations for UFOs—that they were simply misidentified Skyhook balloons. If anyone knew about Skyhook balloons in America, these were the people.

## Project Blue Book and UFOs

In March 1952, the Air Force changed the name of Project Grudge. The Air Defense Command became the autonomous "Aerial Phenomena Group," which was later called Project Blue Book. Ruppelt had an intense aversion to the distorted and distorting term *flying saucer*. To give a more serious tone to the Air Force study, and a more precise definition of what was being studied, Ruppelt introduced the term "Unidentified Flying Object," or UFO.

It should be noted that Ruppelt and his team at Blue Book—in contrast to earlier Air Force attempts to "explain away" sightings—kept an open mind on the subject. Ruppelt made a number of positive contributions. For example, he introduced a standardized UFO questionnaire to make it easier to collect and evaluate sighting data. He appointed Dr. Hynek chief scientific consultant and began the use of cameras that could photograph radarscopes.

Contrary to Air Force and other debunker claims, Ruppelt found that the number of UFO reports *did not* rise and fall in line with the number of newspaper stories about UFOs. It appeared that UFO reports occurred—as difficult as this is for many to believe even today—when ordinary people sighted inexplicable, extraordinary craft in the air.

In the early 1950s, Ruppelt also found that UFO reports seemed to have a relationship to the seasons. For example, there were more reports in the summer months, particularly July. He also discovered that the majority of UFO sightings were around industrial centers, important seaports, and *military installations.*

On April 7, 1952, *Life* magazine stunned the nation by publishing an article entitled, "Have We Visitors from Space?" Hundreds of newspapers nationwide excerpted the article. *Life* appeared to support the idea that intelligent beings from other worlds were visiting Earth. It even seemed possible that top U.S. government officials were using the magazine to "leak" information that they could not come out with directly.

It seemed at the time that officials at the highest level of the American power structure took the idea of extraterrrestrial visitation quite seriously. Major Donald Keyhoe even went as far as to allege that the *Life* article was part of a campaign, approved at the highest levels of political and military leadership, to release information about UFOs and their occupants in such a way that the public would become slowly conditioned to accept this fantastic reality.

Keyhoe was a savvy player in the early days of ufology. He was aware that insiders at the Pentagon or within the Federal government may feed him false information—so-called disinformation. He knew that even the most seemingly helpful information could be designed to mislead and later embarrass him. Keyhoe saw early on that Air Force investigations were designed to squelch serious study of UFOs. He believed the Air Force was determined to make the whole subject of flying saucers an unfit topic for the serious scientist and citizen, to portray the field as the domain of quacks and kooks.

Yet, at times, Keyhoe also thought the Air Force wanted some accurate UFO reports to be published and disseminated. It seemed to him that some of his articles on UFOs were almost "trial balloons" for the Air Force, tests of public reaction to new information. Keyhoe believed that "there's an official policy to let the thing leak out." He believed that his own articles were part of "an education program."

Keyhoe was a very influential figure, and his thoughts helped shape public attitudes regarding UFOs. In addition to expounding on what is now known as "The Extraterrestrial Hypothesis," Keyhoe alerted the public to the fact that UFOs had been visiting the Earth for hundreds of years; that the U.S. government was aware that UFOs were extraterrestrial craft but covered up this knowledge because it feared public panic; and that the aliens controlling the UFOs were observing humanity, especially atomic energy activities, which threatened their safety.

In 1950, the Air Force reported 210 sightings, with 27 unidentified. In 1951, the Air Force received 169 UFO reports; 22 remained unidentified. There were no widespread media reports of UFOs during this period, so the popular press was not influencing the number of reports. By the end of April 1952, after the appearance of the *Life* article, the Air Force had received 82 reports of UFOs. In May 1952, there were 79 more reports. By the end of June, there were over three hundred reports. Then, things quieted down. Some believed the rise in reports could be explained by the *Life* piece on UFOs and felt that when the article faded from the public mind, UFO reports would subside. But it was only the calm before the storm. The curtain was just about to rise on the great UFO flap of 1952.

Project Blue Book data indicate a build up of UFO sightings on the East Coast of the United States in June 1952. In the first three weeks of July 1952, between fifty and sixty reports a week were coming in to the Air Force. Embarrassingly, a good number of these reports were from jet pilots "scrambled" to respond to visual or radar ground sightings of UFOs. By the end of July, the Air Force had recorded more reports than in any single previous year.

At that time, Rupplet conducted a briefing to a group that included the CIA and Naval Intelligence. He reported a disturbing Blue Book finding—*22 percent of the UFO cases were classified "unknown."* Ruppelt felt that these unknowns could be explained away by making a few assumptions. One of the military officers present correctly noted that, by simply making another set of assumptions, one could prove that UFOs were interplanetary craft. According to Peebles, in his book *Watch the Skies!*, the officer asked a perfectly reasonable question, "Why not just simply believe that most people know what they saw?" In 1952, many in the Air Force had concluded that UFOs were spaceships from other planets. The public was unaware of this.

On June 9, 1952, *Life* published a selection of the letters it had received in response to its article, "Have We Visitors from Space?" On June 17, 1952, the government's disinformation campaign began with an article in *Look* magazine by alleged MJ-12 member, Dr. Donald H. Menzel. (Menzel's "double-life" is discussed in Stanton Friedman's book *Top Secret/Majic.*) The article was titled, "The Truth About Flying Saucers," and was written by one of the premier debunkers in the history of ufology. Two weeks later, *Look* published a second piece on UFOs called "Hunt for the Flying Saucers."

By the middle of July 1952, the Project Blue Book staff was innundated with over one hundred solid UFO reports each week. The unidentified sightings had reached a staggering figure—*40 percent*. The East Coast was the epicenter of the "Second American Sighting Wave of 1952."

On three separate occasions—July 10, July 13, and July 14—reputable observers had sighted UFOs in and around Washington, D.C. According to Peebles, in mid-July 1952, an unknown scientist told Ruppelt that in the next few days there would be "the granddaddy of all UFO sightings." The unknown scientists predicted the UFOs would appear over either Washington, D.C., or New York City, adding that he thought it would probably be Washington.

As predicted, a few days later, shortly before midnight on Saturday, July 19, 1952, the UFOs came to Washington, D.C. Who was this scientist and how did he know? On that night at least eight UFOs, and possibly more, appeared over the U.S. capital and were tracked by radar at both Washington National Airport and Andrews Air Force Base.

The UFOs would seem to "drift" along at relatively low speeds and then suddenly zoom off at unbelievable velocities. For hours, experienced pilots and radar operators tracked the UFOs. The mysterious objects penetrated the secure air space over the White House and the Capitol at will. News reporters and photographers jammed the radar room at Washington National Airport. However, representatives of the "free press" were soon ordered out of the area by representatives of the National Security State.

One great fear of these "keepers of the secrets" was that a UFO might be downed by the military jets sent into the skies above Washington. If the media were present at such a momentous, unparalleled event, there would be no way to keep the public from learning that UFOs were truly interplanetary spaceships.

The newspaper headlines captured the futility of the military's efforts to engage the UFOs over Washington: "**Fiery Objects Outrun Jets.**" Out of sheer necessity, the Air Force held a news conference on July 29, 1952—the biggest since WWII—at which Major General John Samford faced the press. "Temperature inversions" were offered as explanations for objects that experienced radar operators had tracked for hours and experienced military pilots had chased in the skies. Once again, UFOs had been explained away. However, weather reports from that night show that the meteorological conditions necessary for temperature inversion to show up as blips on radar screens were not present. In other words, the Air Force had fed the press another false explanation; the press had accepted it uncritically. (A 1969 Air Force study showed that inversions with such characteristics could not even occur anywhere on Earth.)

The Washington Flap—as impressive as it was to the citizens and leaders who witnessed it—did not occur in isolation. America was not singled out for a dramatic display of UFO activity in 1952. As a matter of fact, UFOs appeared all over Planet Earth. And the UFOs appeared in a wide variety of forms—as lights in the night sky, as daylight discs, and even as some close encounters.

## The Secrecy Deepens

In April 1952, UFO sightings rose worldwide. Hundreds of sightings were reported each month. At the height of the global UFO wave of 1952, there were thousands of sightings reported monthly. For example:

- On January 30, a large UFO appeared over **Korea**.
- In **France**, on May 10, a multiple-witness sighting took place at La Roche-sur-Yon.
- On May 31, again in **Korea**, military guards spotted a UFO. A jet fighter was sent to intercept the UFO; a "dogfight" ensued until the object sped away.
- In **Brazil**, a woman reported a fleet of UFOs on June 15. These objects were only *three feet above ground level.* She watched the objects, which were about six hundred feet away, for about thirty minutes.
- On July 1, at Hasselbach, **Germany**, a former Wehrmacht pilot and his son reported seeing not only a flying saucer, but two figures near it wearing shimmering suits. Two other people witnessed the UFO from different locations.
- On July 20, a UFO was reported in **Morocco**.
- From September 20 to 24, NATO naval exercises in the **English Channel** were observed by UFOs. The objects were photographed and chased by military jets.
- In **Morocco**, on September 21, UFOs appeared over Tangier, Casablanca, and Marakesh.
- On September 28, reports came in from all over **northern Europe** and **Scandinavia**.
- In **Australia**, on October 2, two teenagers reported a UFO over Melbourne.
- A fleet of UFOs—along with a "mother ship"—were reported by a schoolmaster and his family at Orlon-Sainte Marie in **France**. These objects were also seen on radar at a nearby station.

- In **Italy**, on November 18, a farmer saw three beings observing him from a UFO that hovered about thirty feet overhead. The beings looked human.
- In the **Gulf of Mexico**, a B-29 bomber reported three UFOs, which were followed by four more UFOs, which were followed by another five UFOs. A huge UFO came into view on radar, all the smaller UFOs headed for it and merged with it. The huge UFO sped off at over 9,000 mph.

Some people can be mistaken some of the time . . . some people can be mistaken all of the time . . . but can *all* of the people of Earth be mistaken *all* of the time?

While flying to Hawaii in April 1952, Secretary of the Navy Dan Kimball saw two objects flying at the unheard of speed of 1,500 mph. Kimball reported the sighting and was outraged to learn that the Air Force would not give him access to their analysis of the report.

A few months later, in July 1952, Navy Chief Warrant Officer Delbert C. Newhouse—a Navy photographer with over twenty years' experience—and his wife were driving near Tremonton, Utah. They saw twelve to fourteen UFOs flying at high speeds in formation. Fortunately (or unfortunately, depending on your point of view), Newhouse had his movie camera with him and was able to capture forty feet of film of the objects in flight. The Navy Photo Interpretation Laboratory near Washington, D.C., studied the footage and concluded that Newhouse had filmed "unknown objects under intelligent control."

At this point, the seriousness of the subject brought on the direct interest of the CIA, and the secrecy deepened. The CIA had been watching the UFO situation carefully from the time the Agency had been created. Admiral Hillenkoetter (another alleged MJ-12 member) told as much to Keyhoe. In November 1952, Republican Dwight D. Eisenhower was elected, succeeding Democratic President Harry S. Truman; Navy Secretary Kimball, who was working to end UFO secrecy, was soon replaced.

In January 1953, the CIA moved to take control of the UFO situation. A secret panel was convened to look at the UFO evidence. At the time, another alleged MJ-12 member was the head of the

CIA—Walter Bedell Smith. Later that year, Smith was replaced by CIA Deputy Director Allen Dulles, the man who had done so much to bring Nazi scientists and psychiatrists to America and to merge the Gehlen Nazi espionage apparatus with the CIA.

Smith had earlier informed the National Security Council in 1947 that about 20 percent of the nearly 2,000 UFO reports could not be explained or explained away. He called for a serious scientific effort to understand the phenomenon, which he believed had national security implications. Smith told the CIA's "Psychological Strategy Board" that UFOs presented serious challenges for U.S. psychological warfare and intelligence operations.

The CIA panel—later known as the Robertson Panel—included another alleged MJ-12 member, Dr. Lloyd Berkner. Dr. J. Allen Hynek, an advisor to Project Blue Book and the Robertson Panel, said that the "truly puzzling cases" were kept from the Robertson Panel. The CIA dismissed the Air Force and Navy evaluation of convincing reports and began to deepen the secrecy surrounding the phenomenon.

The staff of project Blue Book was ordered by the CIA to begin a "debunking campaign" and to plant stories in the press to make the whole subject appear ridiculous. According to Keyhoe, Ruppelt confessed that he could not fight the CIA and would have to go along with their campaign to discredit anyone who made a reliable report and poison the public's mind with misleading and false stories. Deceit was an essential part of the secret campaign.

Although Ruppelt supported the Extraterrestrial Hypothesis in his 1956 book, *The Report on Unidentified Flying Saucers*, he later changed his mind, saying he did not believe UFOs were from outer space. Whether he did so under duress remains a matter of speculation to this day.

While Project Blue Book was represented as the official government investigatory unit for UFO reports, some believed it to be nothing more than a front. There were those who suspected that Blue Book was merely a public relations debunking operation. Was a more powerful group working behind the scenes, taking the important UFO reports and evaluating them in secret? Hynek himself may have believed this; Ruppelt apparently knew that "another intelligence agency" was really in charge of the investigation of UFOs. Who was Ruppelt referring to? We will probably never know.

In his book *Aliens from Space*, Keyhoe wrote that the Air Force had known since 1953 that "giant space ships were operating near our planet." Keyhoe reported that long-range radar had picked up a huge craft orbiting the Earth, 600 miles over the equator, at a speed of 18,000 mph. Similar reports appeared in the 1950s in such publications as *Time* and *Popular Mechanics*. Similar stories were discussed off the record in hushed tones by reporters quoting "reliable sources."

But the great American Iron Curtain of Secrecy had descended. Stories might appear—but how could anyone determine what was a story based on fact from a baseless rumor? How could anyone really know what was accurate or what was false? Who could tell if a source was honest or planting disinformation? In the subterranean world of secrecy, all that lies beneath the mask is . . . another mask. The facts about the UFO phenomenon were effectively kept from the public and from anyone who sought to find out the truth, no matter how high up an individual was in government or the military. To learn what the "keepers of the secrets" knew, you had to have a "need to know." And the clandestine group itself determined who needed to know.

All who enter the morass, the field of quicksand—that is, the investigation of secret government investigations of UFOs— would do well to heed Dante's ancient warning, "All hope abandon, ye who enter here!" Because lying is an essential tool of the culture of secrecy, it is impossible to know without doubt what is completely true, partially true, or totally false when it comes to official involvement with the subject of UFOs.

## Majestic-12

One of the most controversial secret government UFO groups that may or may not have existed is known as Majestic-12 or MJ-12. It is the group that is hinted at in the highly successful television show, *The X-Files,* and in the late (not great) television show, *Dark Skies*. It is the group that rears its ugly head in the excellent films *Starman, Close Encounters of the Third Kind*, and *E.T.*

MJ-12 is the quintessential secret government group. As has been said of other institutions and organizations, if MJ-12 had not existed, it would have been necessary for us to have created it. The clandestine group's membership is so perfect that the individuals involved could have been chosen by the director of central casting—or God.

MJ-12 was a high-level group comprised of top scientists, military leaders, and elite members of the emerging U.S. National Security State. MJ-12 was formed to oversee the analysis of UFO-related information—including that obtained from crashed alien craft—and to keep this information compartmentalized and secret from the American public, from both the allies and foes of the U.S. government and even from most people at the highest echelons of political and military power.

MJ-12 has successfully kept the lid on the most highly guarded and highly classified secrets of this or any other century. Is the U.S. government maintaining a cover-up of its interaction or collusion with extraterrestrials? Is the government hiding truly frightening information? Or is it merely holding back information that would put the government itself in a bad light in one way or another? What the secrets are, and why they have been kept hidden, remain mysteries to this day.

Whether the reader comes to believe that MJ-12 did or did not exist, the best source of information on the subject is provided by Stanton T. Friedman, the author of *Top Secret/Majic* and co-author of *Crash at Corona—The Definitive Study of the Roswell Incident*. Mr. Friedman is not without his critics, but he makes a powerful case for his conclusion that MJ-12 really existed.

Who were the members of MJ-12? In alphabetical order, they were:

> *Lloyd V. Berkner*—a scientist and explorer; worked with Vannevar Bush on the Joint Research and Development Board; a member of the CIA Robertson Panel
>
> *Detlev Bronk*—aviation physiologist; chaired the National Research Council; member of the Atomic Energy Commission medical advisory board; Johns Hopkins, Rockefeller University affiliations
>
> *Vannevar Bush*—organized the National Defense Research Council and the Office of Scientific Research and Development, which created the atom bomb

*James V. Forrestal*—Secretary of the Navy; first Secretary of Defense; replaced by Walter Bedell Smith after his suicide at Bethesda Naval Hospital

*Gordon Gray*—5412 Committee; National Security Advisor to Truman; CIA psychological strategy consultant; Secretary of the Army

*Admiral Roscoe H. Hillenkoetter*—third Director of Central Intelligence; the first Director of the CIA

*Jerome Hunsaker*—MIT-educated aeronautical engineer

*Donald H. Menzel*—astronomer; director of the Harvard College Observatory; high-security CIA and NSA consultant

*Robert M. Montegue*—base commander, Sandia Atomic Energy Commission facility in Albuquerque, NM

*Walter B. Smith*—second director of the CIA; ambassador to the USSR

*Sidney W. Souers*—first director of the CIA; special consultant on security

*Nathan F. Twining*—commander, Air Materiel Command based at Wright-Patterson; helped create Project Sign

*Hoyt S. Vandenberg*—second director of the Central Intelligence Group; destroyed Project Sign UFO report.

This group allegedly came into existence during the Truman Administration and continued to function into the Eisenhower Administration. Many believe that a clandestine group similar to MJ-12 has existed in the United States ever since and has had complete control over all UFO-related information during the past fifty years.

The 1952 wave gave way to the "1957 American UFO wave." Lights in the sky and daytime discs gave way to sightings of ships and even entities near the craft. In the 1960s, the first "abduction" report emerged—the Betty and Barney Hill case.

An earlier 1958 abduction case from Brazil became more widely known. UFO intrusions over military installations in the U.S. continued, and reports came in from NATO, the UK, France, Germany, and the Soviet Union. Another "wave" swept across the United States in 1973. The UFO phenomenon just would not go away.

Rumors of sightings by astronauts grew; stories about sightings in space by members of the U.S. space shuttle crews arose. In 1978, Stanton Friedman opened up the Roswell Incident. In 1980, a major UFO incident occurred in England at the American Bentwaters-Woodbridge nuclear base. In the 1980s, Budd Hopkins presented startling information about the extent of the abduction phenomenon in his books *Missing Time* and *Intruders*. Raymond Fowler extended the boundaries of the UFO phenomenon to include the paranormal in *The Andreasson Affair* and *The Watchers*. In 1992, David Jacobs brought increased clarity to the abduction pheneomenon with his book *Secret Life*. John Mack, M.D., put the abduction phenomenon in a completely different light in his book *Abduction*.

The secret group controlling the emerging UFO information would certainly have its hands full. The UFO phenomenon no longer only involved lights in the sky or daylight discs. The UFO phenomenon was calling into question basic conceptions of reality itself.

After a thorough and painstaking review of the controversial MJ-12 documents, Friedman has concluded that there was without doubt an Operation MJ-12. Other authors vigorously disagree with Friedman and consider the MJ-12 documents to be hoaxes.

According to Friedman, MJ-12 and its successors are sitting on the story of the millenium. He believes that the UFO cover-up amounts to a "Cosmic Watergate," in which a few people know the full story. He would very much like us all to know what MJ-12 discovered about the aliens—where they are from, their means of travel, their reasons for coming to Earth. Friedman wonders what the aliens know about the origin of humanity itself and what Operation MJ-12 has uncovered.

In *Top Secret/Majic*, Friedman poses three questions that he calls the "why" questions:

1. Why has the government covered up the UFO data?

2. Why are the aliens here?

3. Why does it matter that UFOs are visiting Earth and that this information is being kept secret?

According to Friedman, MJ-12 members—an impressive group of human beings—must have been aware of the fact that the aliens were here for their own purposes. In his opinion, after the experience of WWII, this group would have gone to any lengths to keep the crashed Roswell discs and the recovered alien technology completely secret from enemies and friends. Although the Cold War is over, Earth is still not a peaceful planet. Some of the motivations for the original MJ-12 secrecy may be behind the cover-up in effect today by the heirs of MJ-12.

For example, Friedman asks, what would be the impact on major religions if intelligent life from elsewhere were proven and their knowledge challenged the doctrines of a group such as the Catholic Church? How would the national governments of the world react if people responded to the reality of life from elsewhere by abandoning their national allegiances and viewing themselves not as citizens of a particular nation, but as citizens of Planet Earth?

How would the economically privileged—those who worship the new god of "Market Forces" from the top of the financial food chain—react to changes in the economic order produced by new alien technologies that they could not own and exploit? The incorporation of the reality of alien life into human society and consciousness will not be simple and easy, as in a television show or Spielberg movie.

And what if the government knew something horrifying about the aliens? What if there is an apocalyptic truth behind the UFO phenomenon and not a New Age of contact with benevolent Space Brothers. In Friedman's view, all these questions would be faced by the responsible leaders of society today. Continued secrecy may be a temporary necessity in some instances.

Many authors discuss the existence or nonexistence of MJ-12; all are worth reading. In the conflict of information and intuition, each reader can begin to form his or her own informed opinion on the matter. However, it can safely be said that, if UFOs are real, then MJ-12 or a group very much like it would certainly have been created to deal with a phenomenon without precedent in human existence.

The secrecy has deepened over the decades and the cover-up continues. Yet, behind the American Iron Curtain of Secrecy, has there also been an ongoing campaign to bring the whole UFO phenomenon out into the open? Has there been a program to re-educate the American public concerning extraterrestrial life? The evidence for such an endeavor is tantalizing.

**The Mindshift Hypothesis:** Our world has been and is now being visited by advanced intelligent entities from elsewhere, and this reality has been known to a limited number of people within the U.S. government since at least 1947.

# CHAPTER FOUR

# The Mindshift Hypothesis
# The Mindshift Hypothesis
# The Mindshift Hypothesis
# The Mindshift Hypothesis
# The Mindshift Hypothesis

The word "hypothesis" has two generally accepted meanings. According to *The American Heritage Dictionary*, the first meaning assigned to the word is "a tentative explanation that accounts for a set of facts and can be tested by further investigation; a theory." In this chapter, the Mindshift Hypothesis is *not* being offered as either a theory or an explanation for a set of facts. Instead, the word hypothesis is being used in accordance with the second dictionary meaning assigned to it, *"something taken to be true for the purpose of argument or investigation; an assumption."*

The one aspect of the UFO phenomenon that all would agree on—debunkers and true believers alike—is that the phenomenon remains essentially *unknown*. In the face of the unknown, it is not possible to offer an explanation or a theory. It is not possible to take a set of facts about *an unknown phenomenon* and explain them.

However, it is possible to open one's mind, to rid oneself momentarily of prejudices and preconceptions, and to look at the unknown in an entirely new way. It is not easy to do this, but it is possible. One way to free the mind of limiting thoughts is to suspend all judgment and make an intellectual assumption for the purposes of investigation. Detectives do this all the time when trying to solve a crime. Physicians do the same when seeking to make a diagnosis in puzzling situations where clinical information is incomplete or lacking.

The Mindshift Hypothesis *requires that all opinions be set aside.* One cannot have a useful opinion about the unknown. Too often, in science, medicine, and society, opinions parade as knowledge and block deeper understanding and human advancement. The philosopher Nietzsche correctly observed that "opinion is the death of knowledge."

The Mindshift Hypothesis *requires that deeply held beliefs be suspended*, at least temporarily, so that the mind can let new facts stream in and be considered on their own merit. Our deepest beliefs anchor us, providing the stability and equilibrium we need in a world whose only constant is change. But our beliefs can also form a barrier against the perception of new realities, especially against the perception of phenomena that challenge our most cherished beliefs.

The Mindshift Hypothesis *requires that all judgments be delayed* until the facts as they are known have been considered and evaluated. It is critical that each individual not jump to conclusions, whether out of anxiety, impatience, or arrogance. It is essential to get to know the many aspects of the subject under consideration. This takes time when investigating a topic about which much is known; how much more patience is needed when facing the unknown.

Finally, the Mindshift Hypothesis *requires that each individual be honest*—both intellectually and emotionally—with himself or herself. It is difficult to let go of old ways of thinking and feeling that have served us well. This is true for both ordinary citizens and the most brilliant thinkers of any era—past, present, or to come. It is comforting to believe that something is "known" in the great unknown experience we call Life. Most of us are reluctant to give up those ideas and beliefs that keep us going on a daily basis. But, when new realities enter our lives, they can be faced if we are honest with ourselves and accept the facts as they are, not as we wish them to be.

The Mindshift Hypothesis involves making the following assumption: *Our world has been and is now being visited by advanced intelligent entities from elsewhere, and this reality has been known to a limited number of people within the U.S. government since at least 1947.*

If this seems, for whatever reasons, impossible or improbable, suspend judgment temporarily. Each reader is invited to enter into the Mindshift Hypothesis and make the necessary assumption solely for the purposes of investigation. Remember—this is an *assumption*, not a *conclusion*. The hypothesis is a tool that the reader can use to evaluate information whose certainty remains to be determined.

The reader is now faced with an *"If . . . then . . ."* scenario. *If* the hypothesis is correct, and the U.S. government became aware fifty years ago that intelligent beings were visiting our planet, for whatever unknown purposes, *then* what would have been the likely response of the human beings who held responsible positions in government at that time?

Certainly, those in power would have met to discuss the matter. In light of the social and political circumstances of the day, these meetings would most likely have been held in secret. The investigation of life from elsewhere would require the attention of the best minds from all fields—science, psychology, sociology, theology, and, of course, politics and the military. A committee such as the postulated MJ-12 would have almost been a given.

The international conflicts of the day—as well as their own ignorance—would have made those in power eager to keep the knowledge of life from elsewhere secret. All UFO-related information would have been given the highest-level security classification—as indeed it was, in reality, classified at a higher level than even the hydrogen bomb. Those in power would have had to grapple with the uncertain public reaction to this knowledge should it become known to the ordinary American citizen.

In 1900, the U.S. population was only 76 million. Within only thirty years, it had grown to 125 million. By 1947, the U.S. population was about double what it had been half a century earlier. Approximately 50 percent of the population had been born before or just after the Wright brothers flew the first airplane at Kitty Hawk. A large number of people then living in the United States were adults when radio first became a mass communications medium. By 1947, motion pictures had been using sound for only twenty years. That year, television pictures were being successfully transmitted, and a new industry was about to be born.

In 1947, most Americans lived and died within a few miles of their birthplace. Few Americans had direct contact with people from other states, let alone with people from other nations and cultures. The nation had not yet emerged into the postwar boom years; automobiles, leisure time, and traveling for pleasure were still luxuries. The national highway system was nearly a decade away from development, and travel by car was slow and time consuming. Air travel was the privilege of a wealthy few.

Although the American Empire was replacing the British Empire in geopolitics, daily life for the majority of Americans in 1947 was not too different from that of their parents or even grandparents. On the surface, things seemed tranquil. Nazism had been vanquished. Americans cherished democracy, individual freedom, truth and justice, and boasted of what they called "the American Way of Life." (Ironically, the most well-known proponent of the American Way of Life in 1947 was an extraterrestrial—Superman.) To some, a "Pax Americana" seemed at hand.

Although World War II had ended in military victory, global conflict continued. The American war economy grew rapidly. For the first time in its history, the U.S. maintained a standing army in peacetime. The classical political "balance of power" had been replaced by a "balance of terror" because of the atomic bomb. The struggle for racial equality in America was heating up. Jackie Robinson was the first black man allowed to play major league baseball, but ugly racial incidents plagued him on and off the field. Working people fought for advancement and their piece of the American economic pie. Politicians such as Richard Nixon and Joseph McCarthy were building careers based on the exploitation of human fears and weaknesses, which were already poisoning the atmosphere of public discourse.

Those who were aware of the reality of intelligent life beyond our planet faced a volatile world. They faced an America undergoing major social transition, perhaps even transformation. It is possible that those in power could not judge with confidence what the public reaction would be. Or perhaps they knew all too well that such an astounding revelation would result in chaos and serious societal disruption. It may even have been that those in power kept the reality of extraterrestrial life secret to protect their own narrow interests.

Whatever their motivations may have been, it is not surprising that all UFO-related information was kept secret. But is it likely that the sophisticated, educated men who were responsible for guiding society at that time believed that secrecy was all that was needed? Is it conceivable that they thought that the reality of extraterrestrial life could be kept secret forever? Would such secrecy even have been considered desirable?

*If* a small group within the United States government knew about extraterrestrial life, and *if* they knew or feared that the public could not handle that information, *then* what would they do? A program to prepare the American public to be able to accept the reality of extraterrestrial life would seem to be in order. It would most likely have been considered by those in power as a serious option. Could government undertake such a program successfully?

It is fashionable, as this book is being written, to assert that government can do nothing right. It is said that government cannot run social programs successfully, that government cannot formulate economic policy effectively, and so forth. Yet, effective government programs such as Medicare and Social Security provide much needed service every day to tens of millions of people. And government economic policies, such as the very popular home mortgage interest tax deduction, are hailed by millions of people who rail against government promotion of a social agenda through the tax code. So it seems that government can do many things effectively.

It is also fashionable to claim that government cannot keep secrets. Such an assertion would be laughable to the individuals who are now working day and night to declassify 190 million pages of formerly classified government documents on a variety of subjects. Executive Order 12958, signed into law by President Clinton in 1995, requires that massive amounts of secret documents be declassified in the next few years.

Many do not believe that government can conduct an effective propaganda or educational campaign. In fact, the U.S. government ran a successful domestic propaganda campaign in the years leading up to World War II. Isolationist sentiment in America was strong, and Americans were inclined to remain neutral and stay out of Europe's conflicts. In addition, there was also a significant segment of the American population—at all economic levels—who admired Adolf Hitler and would have supported an American brand of fascism if given the opportunity. The novel *It Can't Happen Here*, by Nobel prize–winning author Sinclair Lewis, portrayed the rise of fascism in America. And the film *Meet John Doe*, by Oscar-winning director Frank Capra, told the story of an attempt by American corporate fascists to seize power in the United States. (President Roosevelt later put Capra in charge of the U.S. film propaganda program as part of the war effort.) Under the Roosevelt Administration, the U.S. government used every tool at its disposal to prepare the public for the necessity of entering and winning the war in Europe. And it did so successfully.

Another successful government campaign, involving both secrecy and propaganda, began in the late 1940s and continues to this day. The promotion of atomic power—for both military and civilian uses—required the utmost secrecy and an extraordinarily effective public relations propaganda campaign. This is detailed in such books as *The Cult of the Atom* by Daniel Ford and *The American Atom—A Documentary History of Nuclear Policies from the Discovery of Fission to the Present*, edited by Philip L. Cantelon, Richard G. Hewlett, and Robert C. Williams.

From the beginning, the Atomic Energy Commission had a dual purpose: first, to promote atomic power to benefit the American atomic energy industry; and, second, to protect the American public from the hazards of atomic energy.

Unfortunately, the interests of the atomic energy industry invariably came before those of the American people. The government propaganda program misled the public into believing that atomic energy would produce electricity that would be "too cheap to monitor." Today, nuclear power plants are too expensive to run and maintain. All across the United States, over one hundred civilian nuclear power plants hang over our heads like radioactive Swords of Damocles—each one deteriorating according to the known laws of nuclear physics, each one a future Chernobyl.

However, the successful government campaign of secrecy and propaganda lulled the American public into complacent acceptance of "safe, clean, and efficient" nuclear power plants that later proved to be unsafe, toxic, and inefficient. Although some shocking secrets concerning government human radiation experimentation were released by President Clinton's Secretary of Energy Hazel O'Leary, the darkest truths about atomic energy remain closely guarded to this day.

It seems, therefore, quite possible for the government to maintain a program of secrecy concerning UFOs and extraterrestrials. If the truth about nuclear energy can be kept secret for fifty years or more, why not the truth about extraterrestrial life? It seems equally possible that the government could successfully execute a propaganda or educational campaign directed at the public concerning the reality of intelligent alien life. If the government can carry out a fifty-year propaganda campaign for a deadly technology, can it not also devise a program that would help the public gradually accept the reality of extraterrestrial life?

## Exploration of a Program of Public Re-education

Just as accepting the possibility of alien life is not identical to accepting the idea that extraterrestrials are visiting earth, believing that the government *could* carry out a campaign to re-educate the American public concerning extraterrestrial life is not the same as believing that it *did*.

In fact, even those who accept the existence of MJ-12, or a supersecret organization like MJ-12 that would have been responsible for such an effort, do not necessarily believe that there was a campaign to re-educate the public concerning extraterrestrial life. For example, Stanton Friedman is one of the staunchest proponents of the existence of MJ-12, yet he does not think there was a government education campaign. Apollo astronaut Edgar Mitchell also believes that MJ-12 existed, but he knows of no evidence indicating a formal government education campaign regarding extraterrestrial life.

On the other hand, noted ufologist Raymond Fowler—whose most recent book is *The Andreasson Legacy*—sometimes feels that such a program does exist. Michael Lindemann, a talented journalist and publisher of the electronic newsletter *CNI News*, believes that, if people in government know that extraterrestrials are visiting earth, then they would be *obliged* to develop such an education campaign.

In 1997, a controversial book was published by an equally controversial author—*The Day After Roswell* by Lieutenant Colonel Philip J. Corso, a retired Army Intelligence officer. As is to be expected, Corso has his defenders and his detractors. Some ufologists who have read his book and interviewed him think that, overall, his story has the ring of truth. Other knowledgeable people believe that Corso has exaggerated the truth, thereby doing a disservice to those trying to get to the facts. Still others assert that there is no evidence to support either the details of Corso's story in general or his description of the role he played in the secret military "reverse-engineering" of alien technology in particular. It is in just such situations as this—when the experts disagree on a significant subject—that each person's intuition may be more important than information.

In his book, Corso openly discusses what is being referred to here as the Mindshift Hypothesis. Corso claims to have played a role in getting advanced German rocket technology out of Europe through Italy. He correctly describes the American and Russian exploitation of Nazi science and Nazi scientists at the end of World War II. According to Corso, those who learned of the crashed UFO at Roswell sought to keep two unprecedented secrets: the fact of the UFO crash itself; and the beginning of an American research and development program designed to "reverse-engineer" alien technology.

The decision was made at the highest levels to keep the American public and other nations in the dark about these two subjects. President Truman formed a "working group" to take control of UFO-related affairs. This group was comprised of the same individuals that Stanton Friedman has identified as being members of MJ-12. President Truman and the working group came to the conclusion that hiding the UFO story over the long run was impossible. As they pondered how best to keep UFO reality secret, those in power watched in amazement as the national press—all on its own—downplayed, ridiculed, or simply failed to cover the UFO phenomenon. The working group could not have demanded the cooperation it was receiving from the free press.

Corso reports that the working group puzzled over the problem of how to collect good data on the UFO phenomenon while, at the same time, keeping a lid on the subject and preventing anyone from thinking UFOs were taken seriously by people in power. According to Corso, Lt. Gen. Nathan Twining suggested that cover stories be released to explain away UFO sightings while the working group actually performed critical research covertly.

In addition, Corso states that Twining emphasized that the public needed to be prepared for eventual disclosure of the full truth about UFOs and their intelligent extraterrestrial occupants. In the view of the working group, the public needed to be gradually "desensitized." The group feared that panic would be the result if the truth about alien life were made known suddenly to an unsuspecting and unprepared populace.

In his book, Corso reports that Twining believed that the activities of the working group would entail the creation of both an extensive cover-up of the story of the century and, simultaneously, the gradual release of the UFO information being covered up. Corso states clearly that a secret government, hidden from the constitutional government, was formed. Apparently this invisible government was entirely independent of all duly elected officials of either major party.

While the working group steadfastly maintained its grip on all UFO-related information, it also released some extraordinary material. In keeping with its dual policy of both revealing and concealing information, some of the material released or leaked was true, some of it was false. Corso informs the reader that all information, regardless of its accuracy, was used to help create a climate in which the idea of alien life would become less frightening.

In dialogue that could come from an episode of *The X-Files*, Corso quotes Twining as saying, *". . . the cover-up is the disclosure and the disclosure is the cover-up."* The working group came to believe that in this way—with true, false, and partially true information all mixed together and released—the public and the media would be confused but would continue to speculate on the subject of UFOs. Over time, the reality of life beyond Earth would become more acceptable.

In the Afterword to his book, Corso addresses the question of government's ability to keep important information secret for long periods of time. He describes an ongoing project, codenamed "Shamrock," in which the U.S. government convinced the communications giants—ITT, AT&T, Western Union, and RCA—

to allow the U.S. military to eavesdrop on all international calls—for national security purposes, of course. Corso states that Shamrock continued to operate covertly for twenty-eight years, from 1947 through 1975, and was unknown to presidents Eisenhower, Kennedy, Johnson, Nixon, and Ford. Corso believes it is entirely possible for one part of government to keep secrets from even the highest levels of political authority.

If Corso is correct, is it not reasonable to consider whether or not his book is a part of the government campaign he describes? Is it not likely that—as a part of this campaign—Corso's book is a mix of accurate information, complete falsehood, and partial truth? Is *The Day After Roswell* itself a part of the ongoing effort to re-educate the public? It is crucial to consider the source of information and to look at the facts presented from all angles when reading about a complex, poorly understood subject such as the UFO phenomenon.

For the working group, the secret campaign to desensitize the public would be carried out over generations. Over time, the generation born before World War I would pass away. The generation born before the Great Depression would pass away as well. And the generation of the working group would itself pass away. But the covert campaign to re-educate and desensitize the public would continue on.

Between 1947 and 1964, during the "baby boom," approximately 75 million people were born in the United States. By 1990, more than 50 percent of the U.S. population had been born since July 16, 1969, the day American astronauts landed on the Moon. In 1998, there are about 265 million people in the United States. All during this time, the "working group" of Corso, or the "MJ-12" of Stanton Friedman, or the "clandestine group" that many in ufology speculate has long existed, continued its educational effort and its secret research.

This research has included the study of the biggest find of all—a captured EBE, or Extraterrestrial Biological Entity. The secret group also has studied the technology retrieved from crashed alien spaceships. Corso claims to have played a major role in "seeding" American industry with bits and pieces of alien technology, which were then eventually developed into such familiar features of modern-day life as integrated circuit chips, fiber optics, lasers, and super-tenacity fibers.

Not everyone in ufology accepts Corso's description of how this seeding and reverse-engineering took place and his explanation of his role in the process. However, some highly knowledgeable people, who reject Corso out of hand, confirm the fact that human beings have indeed taken alien technology from crashed spaceships and back-engineered or reverse-engineered the extraterrestrial technology into such miracles as lasers and computer chips.

Is Corso alone in his astounding views on secret government activity regarding UFOs? In 1997, another fascinating book was published—*UFO* by Charles Sellier. Sellier is a writer and film producer with over thirty feature films and seventy-seven made-for-television movies to his credit. Early on in his book, Sellier describes his own experience with government censorship and the lengths to which it will go when enforcing secrecy about UFO-related information.

In 1979, Sellier began work on a nonfiction book entitled *Hangar 18*. In UFO lore, Hangar 18 is where the crashed Roswell alien disc was taken to be studied by the military. In addition to his book, Sellier planned to make a documentary film based on the same information. As soon as word of the film was released, Sellier and his staff began receiving unusual requests for specifics about the project from unknown sources. Later, they learned that the Department of Defense was trying to get a copy of the script indirectly from other sources.

Rumors reached Sellier that the Department of Defense was applying pressure behind the scenes to prevent his film from being made. The government's efforts were successful. Sellier was called to a board meeting of the parent company of the Schick Electric Corporation and told that he could only make his film if it were done as *fiction*. He could not make the film as a documentary.

The actions of the Department of Defense may lead one to wonder, "If the government can suppress films such as Sellier's, can it also support projects, plant ideas, or cooperate behind the scenes in the production of films and television shows that promote its views?" Such activity would be perfectly in line with the goals of a group determined to reshape public attitudes concerning UFOs. It would not require a massive effort on the part of large numbers of people. One project could be suggested here, another one planted there a few years later. The economics of the entertainment business would do the rest.

According to Sellier, there is evidence that such intrusion by the government dates back to the Truman Administration. Sellier states that he believes there is reliable information indicating that MJ-12 exists today in a new form and that its secret code name is PI-40. Sellier wonders if this top-secret group, and perhaps its counterparts in other nations as well, have already made contact with extraterrestrials. In his book, Sellier suggests that even the most outlandish information regarding UFOs may deserve serious consideration because almost everything remains to be learned.

## Assessment of the Situation

Who is visiting our planet? Why are they here? Is there a plan? Are these visitors benefactors, foes, or completely indifferent to mankind?

In 1948, the Air Force's secret UFO group, Project Sign, prepared a report called the "Estimate of the Situation." At that time, the Air Force top brass wanted to explain away all sightings and dampen public enthusiasm for flying saucers. However, this "Estimate" advanced the so-called "Extraterrestrial Hypothesis." It was immediately shot down and was later ordered destroyed. However, the very destruction of the "Estimate" fostered belief in its conclusions in the minds of many. The "Estimate" remains an influential and controversial UFO document to this day.

In 1963, another controversial UFO document came on the scene. At that time, a young man, Master Sergeant Robert O. Dean, received NATO's highest security clearance, "Cosmic—Top Secret." Dean was given access to a report that appeared to substantiate and confirm the "Extraterrestrial Hypothesis." Simply called "Assessment," the document allegedly concluded that:

1. At least four alien civilizations are conducting massive investigations of our planet. One species of alien could pass for human, undetected by anyone.

2. These extraterrestrials are far superior to us, with technology thousands of years ahead of ours.

3. Alien activity is not random; there seems to be a definite plan.

4. Extraterrestrials have been visiting our planet for many, many thousands of years.

5. There is no apparent military threat.

According to Dean, these were the conclusions of the scientists, theologians, psychologists, and social scientists who contributed to the "Assessment." Other UFO researchers, such as Linda Moulton Howe, have reported similar information that appears to be in line with Dean's claims.

Although Dean is breaking his secrecy oath and is violating the National Security Act, he is allowed to speak his mind freely, to lecture and write on the subject, and to appear on television and in video productions about UFOs. Why? Are Dean's public comments part of the *"disclosure that is the cover-up"* that Corso suggests exists? Dean would not even have to be an active part of the covert activities for this to be true. In fact, it would be more effective if he were truly independent. Are ideas such as this allowed to enter public consciousness as part of the re-education campaign?

Sellier writes that, in Dean's opinion, the aliens have somehow played a crucial role in our very existence on Earth. Furthermore, Dean believes that the aliens are now in the process of asking us to join with them. But first, it seems we must change our behavior and become a less violent species. In addition, we must let go of old ways of thinking about who we are, about the nature of Life, and about our role in the Universe. In other words, humanity needs to undergo a *mindshift* before it can take its place in the cosmic community. It remains to be seen if the people of Earth are up to the challenge.

Today, UFOs and extraterrestrials are everywhere—in the movies; on prime time television comedies and dramas; in popular music; and in print, radio, and television advertising campaigns. In 1996, Republican presidential candidate Bob Dole referred to Roswell in stump speeches. In Northern Ireland, President Clinton was asked if a UFO really did crash at Roswell. At first glance, it would seem that the re-education campaign had, after fifty years, achieved its goal of desensitizing the public to the reality of alien life.

But the cover-up and the secrecy continue. Why?

Dr. Steven M. Greer, founder of the Center for the Study of Extraterrestrial Intelligence (CSETI), and his colleagues have identified and located several dozen important government witnesses whose stories would make the case for the existence of extraterrestrial lifeforms undeniable. However, all of these individuals are restrained by national security oaths and other security restrictions. They cannot tell what they know.

In addition, Dr. Greer and other members of his Project Starlight Team, have met with senior members of President Clinton's Administration, Congress, the CIA, and the Joint Chiefs of Staff. Apparently, most senior levels of political and military leadership in the United States are out of the loop regarding UFO information. It seems that UFO-related information is in the hands of an extra-constitutional group functioning outside the normal channels of government. The UFO cover-up is obviously still continuing.

It is interesting to note that, while the government officially denies the existence of alien life and extraterrestrial spaceships, the United States Navy has a "recommended UFO reading list" available to its personnel through the Naval Historical Center in Washington, D.C. Among the titles listed in the Navy's UFO bibliography are:

- *Flying Saucers, Farewell*, by George Adamski, a famous 1950s contactee
- *The Roswell Incident*, by Charles Berlitz
- *The Day After Roswell*, by Philip J. Corso
- *Crash at Corona: The U.S. Military Retrieval and Cover-Up of a UFO*, by Stanton T. Friedman and Don Berliner
- *Flying Saucers from Outer Space*, by Major Donald E. Keyhoe
- *Abduction: Human Encounters with Aliens*, by John Mack, M.D.
- *A History of UFO Crashes*, by Kevin D. Randle
- *UFOs Are Real: Here's the Proof*, by Edward Walters and Bruce Maccabee

The bibliography contains many other interesting books and also includes the complete record of Project Blue Book, many comprehensive UFO bibliographies, twenty reels of UFO motion pictures received by the Air Force between 1950 and 1967, and a number of reprinted articles about UFOs from the publication *Soviet Soldier.*

So many questions remain. Does the Mindshift Hypothesis describe an ongoing reality? Is the Navy bibliography a part of the campaign to gradually re-educate the American public about extraterrestrial life? If there is nothing to the whole UFO business, as the government claims, why are witnesses held to forty- or fifty-year-old security oaths that prevent them from telling what they know about UFOs and their occupants?

The Mindshift Hypothesis is a useful tool with which to evaluate the UFO phenomenon. But unrelated to any postulated government program, independent, hardworking UFO researchers have been investigating this subject for decades. They have helped create a mindshift regarding aliens and extraterrestrial spaceships. What do the pioneers in the field of UFO research think about the Mindshift Hypothesis and other crucial but unanswered questions concerning extraterrestrial life?

Dr. Edgar Mitchell

Stanton Friedman

Budd Hopkins

Dr. John E. Mack

Raymond Fowler

Dr. David Jacobs

Don Berliner

Michael Lindemann

# CHAPTER FIVE

# The Mind-Shifters
# The Mind-Shifters
# The Mind-Shifters
# The Mind-Shifters
# The Mind-Shifters

The New York Public Library is one of the finest libraries in the world today, offering access to the wisdom of the ages to people from all walks of life. Its main branch on Fifth Avenue in Manhattan is home to a collection of over 500 books on UFOs, flying saucers, and extraterrestrials, about 350 of which have been published since 1972. As impressive as this collection is, it is only the tip of the iceberg.

Books exploring the subject of UFOs have been written by men and women from the United States, the United Kingdom, the former Soviet Union, France, Germany, Italy, Brazil, South Africa, Japan, China, and many other nations all over the world. Some are first-person accounts of UFO sightings or abduction experiences; others are written by scientists and engineers addressing the technological implications of UFOs. Some are popular overviews of the subject; others are more authoritative, even scholarly, approaches to the UFO phenomenon.

In their own way, each of the authors has made a contribution to ufology, providing another piece of the puzzle. Yet some authors have captured the public's attention and imagination more than others. Some UFO researchers have changed ufology dramatically and helped alter public thinking on the subject. In the 1940s and 1950s, Major Donald Keyhoe's books and magazine articles shaped public perception about UFOs for a generation. In the 1960s, John Fuller's article in *Life* magazine, and his book *The Interrupted Journey*, introduced the phenomenon of alien abduction to a mass audience through the story of Betty and Barney Hill.

In a similar fashion, over the past three decades, the UFO researchers, investigators, and journalists the reader will meet in this chapter have all made valuable contributions to our understanding of the UFO phenomenon, who we are, and what our place in the Universe may be. By way of introduction, they are:

***Apollo 14 Astronaut, Edgar Mitchell, Sc.D.,*** author of *The Way of the Explorer* and *Psychic Exploration: A Challenge for Science*, is one of only twelve human beings to have walked on another heavenly body. On his way home from the Moon, Dr. Mitchell saw the Earth from the perspective of an approaching extraterrestrial. He underwent a transformative experience that changed his life. His outer journey to explore the Moon led to an inner journey and the scientific exploration of the phenomenon of consciousness.

***Stanton T. Friedman,*** co-author of *Crash at Corona* and author of *Top Secret/Majic*, is another scientist whose work led him to investigate the UFO phenomenon. Friedman, a former nuclear physicist, was the man who got the first Roswell witnesses to speak about their experiences. He is also the leading expert on the ultrasecret government UFO group, MJ-12.

***Budd Hopkins,*** through his bestselling books *Missing Time*, *Intruders* and *Witnessed*, has made "UFO abduction" and "abductee" household words. His books have opened up an exciting—although sometimes frightening—new chapter in the UFO story.

***John E. Mack, M.D.,*** author of the best-selling book *Abduction— Human Encounters with Aliens*, is an internationally-renowned Harvard psychiatrist and Pulitzer prize–winning biographer of T. E. Lawrence (the famous Lawrence of Arabia). His abduction work with hundreds of "experiencers" nearly cost him his job and his reputation.

**Raymond Fowler** has been a major contributor to the field of ufology for over thirty years. He is the author of nine books on UFOs, among them *The Watchers, Parts I and II; The Allagash Abductions; The Andreasson Affair; The Andreasson Affair— Phase Two;* and *The Andreasson Legacy*.

**David Jacobs, Ph.D.,** author of *The UFO Controversy in America, Secret Life,* and *The Threat,* has studied the UFO phenomenon for over thirty years. He teaches the only official course on UFOs in the United States at Temple University. His latest book, *The Threat,* advances some unsettling ideas about an "alien agenda."

**Don Berliner** is the co-author of *Crash at Corona* and the primary author of the controversial, limited-edition report, *Unidentified Flying Objects Briefing Document: The Best Available Evidence,* which was funded by Laurence S. Rockefeller. In February 1998, he was elected Chairman of the Fund for UFO Research.

**Michael Lindemann** is the author of *UFOs and the Alien Presence—Six Viewpoints* and publisher of the electronic newsletter *CNI News.* He is a superb journalist who has been investigating the UFO phenomenon with an unbiased eye since he woke up to UFO reality in August 1989.

Although each of these investigators has developed a distinct viewpoint on UFOs and extraterrestrial life, none claim to have *the* answer or anything remotely approaching a comprehensive understanding of the situation. However, when taken collectively, their insights advance our knowledge of this perplexing phenomenon.

## The Dark Side of the Moon

Dr. Edgar Mitchell is a graduate of MIT, with a doctorate in aeronautics and astronautics. As an astronaut, he trained for three missions—Apollo 10, 14, and 16. On Apollo 14, Dr. Mitchell flew as the lunar module pilot. On January 31, 1971, Apollo 14 lifted off from Cape Kennedy in Florida. Three days later, Edgar Mitchell and Alan Shepard walked on the surface of the Moon.

He is one of only twelve human beings to have done so.

As he traveled through space, heading home to Earth, Dr. Mitchell experienced a profound sensation that changed him utterly. He felt an overwhelming sense of a universal connectedness. He sensed the *interconnectedness* of all things—himself, his fellow astronauts, Planet Earth, and the vast cosmos itself. Dr. Mitchell sensed that the cosmos was in some way conscious.

In 1972 Dr. Mitchell founded the Institute of Noetic Sciences—an organization he initially envisioned to be more of a *state of mind* than a physical entity—to allow him to function as an independent scholar and pursue his research into the unknown area of consciousness. He investigated both the mystical viewpoint, which assumes consciousness to be a primary phenomenon, and the scientific viewpoint, which assumes consciousness to be secondary. The scientific explorer of the outer world became a scientific explorer of the inner world. His journey of over twenty-five years has taken him to a subject of inquiry he resisted at first—UFOs.

Dr. Mitchell began his remarks to me on a cautionary note.

"Contrary to popular reports, we had no experience *in the astronaut program* that ever came close to what you would call a UFO experience. Of course, we saw things that you could not immediately identify. But there were no *confirmed* sightings in space by any of the American astronauts or, as far as I know, by any of the Soviet cosmonauts that amount to UFO sightings in space, while I was with NASA. However, the persistence of these reports led me to investigate further.

"Since then, I have had the opportunity to speak and work with *military, government, and security* people who claim that—as a result of their positions and their jobs—they have had some sort of UFO experiences. And they are prepared to tell their stories, even though most of them operate under high security clearances."

Dr. Mitchell reemphasized that he was not speaking about UFOs from personal experience. "The idea of considering whether UFO reports had any validity became a very important question. The persistence of these reports led me to investigate further. I knew Dr. J. Allen Hynek, who was one of the leading early investigators of UFOs and know many current professional investigators. I can only speak from the experiences of credible observers. There are numerous reports from average human beings that cause you to think that there must be something to the UFO phenomenon. It's not that I don't value the experiences of the average person. I do. But I tend to rely on those whose positions in government were devoted to this sort of phenomenon, the UFO phenomenon. *And they tell me that there has been alien presence.* I tend to believe their testimony."

As a highly credible individual himself, Dr. Mitchell's observations and thoughts must be considered seriously. "It is a totally antiquated idea—theologically and scientifically—to believe that we are the only intelligent beings in the universe. However, until twenty-five to thirty years ago, this was the accepted wisdom. Since then, we have begun to disabuse ourselves of this idea and recognize that it's the height of foolishness to think we are alone in the Universe."

In 1996 and 1997 alone, newspapers and magazines, radio and television exploded with exciting news about ancient microbial life on a Martian rock; the possibility of life on Mars itself; the possibility of life on the moons of Jupiter; and the potential for water on the Moon, a prospect with vast implications. Almost weekly, astronomers seemed to find planets capable of supporting life revolving around other stars.

Dr. Mitchell continued, "We don't really have, *in the public domain,* 'smoking gun' evidence that we can put our hands on that shows we have been contacted by extraterrestrial civilizations. However, I tend to think that there is evidence *not in the public domain* that points in that direction. In the public domain, we do not have evidence that is available and that can be verified that indicates that intelligent life is visiting Earth. But my informants, the people I talk to, claim that such evidence is available and that it is held by *a clandestine group.* The evidence of ET reality is there. It's just not available, either to the public or to high-level political leadership."

Many in ufology have speculated and argued over the existence or nonexistence of a clandestine group operating behind the scenes. This book itself makes the assumption that some such group is in existence. Dr. Mitchell speaks with the authority of his own research and experience on this topic. As controversial as some of his statements may seem at first, it behooves us all to consider them with an open mind.

"MJ-12 did exist. It did exist. And there is some indication that an organization like MJ-12—maybe under a different name—*is still operating.* There has been confirmation, or at least partial confirmation, that what I am saying is true. It seems that by the time of the Eisenhower Administration, very few high-level political or military leaders were let in on the UFO information. This tended to produce, from all the accounts I have been able to put together, a kind of *secondary or clandestine-type organization within the U.S. government itself* that was running the show."

According to Dr. Mitchell, there was a valid reason for secrecy concerning UFOs in the early days. "We were barely out of World War II then and engaged in the Cold War when the modern period of UFOs began. And it wasn't clear what these objects were. Could they possibly have been German or Soviet equipment? If they were as exotic as they seemed, there was cause for alarm because the military had no defense against them. It would be a perfectly valid rationale, in time of war, for keeping things quiet. There was also the fear of public panic, such as ensued after the Orson Welles's radio broadcast of *War of the Worlds* in 1938. I think there were good reasons for secrecy at that time.

"Unfortunately, government secrecy seems to have gotten out of control. UFO information seems to have been classified at such a high level that even top-level political and top-level military leadership was *systematically* excluded because they were not considered to have a 'need to know' about these things. I think that this has become dangerous.

"The records from the early period have been simply removed, or disposed of, or secreted in such a way by whomever is doing it, so that they are not available today to people in government. We don't know whether they have been destroyed or are kept inaccessible."

In Dr. Mitchell's view, the secret activities of the clandestine group are not in keeping with our Constitutional form of government. In fact, such activities may pose a grave danger to our democracy. He believes it is possible for the government to be more forthcoming about UFO information from the earlier period of history. "It's still possible for the government to be more open. The problem is that the records from that period are simply not available in open government."

To Dr. Mitchell, the preponderance of the evidence of UFO reality and extraterrestrial activity on Earth is convincing. There is no one individual story or incident that is definitive. It is the corroboration that one story provides for the other that gives the ring of truth. "No one UFO story alone gives you an absolutely firm grip on this. But all of these stories point like arrows to a solid body of fact which, unfortunately, we can't completely validate yet. But we are getting people with a certain amount of credibility who can be checked out reasonably well. They were doing what they said they were doing, at the time they said they were doing it. This gives credibility to their stories. But it doesn't make them fact.

"We are reaching a point at which there are so many sightings, and so much circumstantial evidence recovered, that there is no doubt, *utterly no doubt whatsoever,* that strange flying machines and strange technologies beyond the normal arsenal of constitutional government are operating. The question remains: who is operating them? To answer that question, we need to get to the people who have the evidence—while they are still alive and able to come forth and talk about it."

And who is operating the exotic craft that are spotted increasingly in the skies over America? It may not be a simple question to answer. Dr. Mitchell says, "I think one would make a mistake if one concluded it is all ET activity or that there was no ET activity. You have to allow for the possibility of both. I can't speak for some of the private groups I work with on this. There are several I am not at liberty to name. But I have consulted with Steven Greer, M.D., of CSETI. And I have worked some with John Mack, M.D. Dr. Greer has spent several years, using his own professional resources, following leads, trying to get a handle on this. It is my personal bias that some of it could very well be ET, but that some of it is not. We are dealing with a very complex set of problems here."

Among the many stories and rumors one hears about ET activity are the tales of crashed extraterrestrial spaceships. Most people in the United States today have heard of the Roswell story. In the book *Crash at Corona,* Friedman and Berliner write of *two* crashed spaceships. And Dr. Greer, on his CSETI website, has a list of *over 155 UFOs that may have crashed* since the 1940s. Also, in relation to crashed spaceships, stories of recovered alien technology have circulated for years. *The Day After Roswell* by Philip J. Corso is the most recent book to bring reports of alien technology to the public's attention.

Dr. Mitchell has heard the stories as well. In November 1997, at a conference in New York City sponsored by the Friends of the Institute of Noetic Sciences, Dr. Mitchell was asked about such things as back-engineered alien technology and the observation of alien bodies. "People with official positions in government who knew of the data, the materiel, the bodies—these folks are telling the story. But they are under high security classification. It is long past time for the public to know. I give their stories ninety-nine percent credibility." Dr. Mitchell again emphasized that UFO information needs to be brought under congressional oversight and made accessible to high-level government officials.

Dr. Mitchell believes that there has been reverse-engineering of alien technology. "Unfortunately, I think Corso's book has been compromised. And that's quite a shame. Some of the things he claims in his book most authorities would say are not true. He *overstated* the case. For some of the technology involved, we know the line of development and this technology was not reverse-engineered. This casts doubt on Corso's credibility. However, people I know from the group that I work with conducted personal interviews with Corso. They say he didn't make some of the overstated claims in the interviews. The claims that he made to our interviewers are substantiable. They seem to be solid. But the book overstated the case; that's what I mean by compromised. There could be some disinformation involved, but we don't know the answer to that."

To Dr. Mitchell, humanity's view of itself and its place in the Universe has broadened considerably since the advent of the Space Age. Educated, alive, thinking, searching people everywhere are changing as we learn and understand more about ourselves and the cosmos. Dr. Mitchell observes, "As the evidence mounts, and we understand more about the nature of the Universe, it's a preposterous notion to think that we are unique and that this little planet is the only planet that has intelligent lifeforms capable of managing their environment. It is clear that this is an evolutionary concept that we are dealing with. It's evolutionary in this sense: whatever pertains here undoubtedly pertains elsewhere in millions—if not billions—of other places in the Universe. This is a concept that has only really come to be of significance in the last twenty-five to thirty years. Although many of us have held this view for longer than that, it has only recently become accepted common knowledge that the Universe is structured in this way. That means a changing paradigm in science, religion, and culture."

The UFO phenomenon has invaded our earthly reality. Although it is not yet understood—or perhaps precisely because it is not understood—this unexplained reality is forcing the human mind to stretch, to question, to search. The UFO phenomenon is not simply about lights in the sky, daylight discs, or alien abductions. It touches on fundamental questions of our existence and the nature of the Universe.

In his book, *The Way of the Explorer,* Dr. Mitchell discusses what he calls the Dyadic Model, a way to deal with knowledge, existence, and our knowledge of our existence. "The previous notion in science was that the Universe was a giant accident and that it would be too much of an accident for life to have formed, and become intelligent life, in more than one place at a time. To science, that would be too much of an accident.

"That view is clearly wrong and science has had to revise its notion and accept the fact that whatever process has brought us into being has brought life into being throughout the Universe. Now that we look back on it, it sounds silly that throughout our history, until the last two or three decades, we really believed we were alone in the Universe. Now that we see beyond that, we see how archaic these types of ideas are.

"My work is attempting to deepen our understanding of the mechanisms by which all of this took place. It is an attempt to bring together the various claims of science and the various religious cosmologies to understand that there are a set of principles operating here. These are evolutionary principles—*processes*—by which the Universe not only evolved itself, but also, by which the complex process we call 'knowing' or 'consciousness' can be fit into a common model. Physicists have had enough time to bring together quantum physics and Einstein's general theory of relativity—the two most powerful physical theories of this century. Now we add such questions as 'How did we become conscious? What is the process? What is the very nature of our *knowing* about our existence? How did that come about?'

"We return to these two old models—science and religion. Science says consciousness is just an accidental process. The cultural-religious model says that consciousness is fundamental. Accidental organization of energy-matter to become conscious is one extreme; Deity is the other. Neither of these are viable options. The questions still remain: Who are we? How did we get here? How did all this come about? We need a good, overarching cosmological model that explains our existence—both how we are here and how the amazing awareness we experience comes about.

"The model that most of us prefer to use is not a Cartesian-type model in which mentality and physicality are two different things. They both must fit within a common model. This means that *energy* is clearly the fundamental concept. By energy, we mean an action principle, that which causes things to happen. The difficulty here is the interaction between knowing and existing; between intentionality and perception. How all that interacts with physicality.

"Most of us are aware today that *energy has become aware of itself.* That's a nice statement but the real question is, how? How do we measure the energy? How do we know about the energies? How do they relate to each other? All of these physical phenomena we experience—such as energy fields around the body—what type of fields are they? Are they of the type of energy that are offshoots and combinations of the electromagnetic structure that we know about? Or is the energy really something different that we don't know about?

"I, and many of us, take the view that the only real energy that we know about is part of the electromagnetic spectrum. That vibration and movement and energy are all interrelated. We are probably not dealing with strange types of energies that we don't know about, although we certainly deal with manifestations of energies that we haven't properly codified."

What might be an example of a type of energy that science has not yet codified? In Chapter 11, the reader will be introduced to the work of the scientist Wilhelm Reich; in particular, his thoughts on extraterrestrial life. His concept of "orgone energy" is a good example of such an "uncodified" energy. In *Cosmic Superimposition*, published in 1951, Reich wrote, *"The quest for knowledge expresses desperate attempts at times, on the part* *of the orgone energy within the living organism to comprehend itself, to become conscious of itself. . . . Here we touch upon the greatest riddle of life, the function of* **self-perception** *and* **self-awareness**.*" Does this relate to Dr. Mitchell's work?

Dr. Mitchell said, "I am very familiar with Reich's work. But as for orgone energy, I don't know what that is if it is not within the electromagnetic spectrum. If it can be described as a movement within a medium, we know how to describe that as 'wave form.' It seems that the term wave form is the proper way to describe the energy that we are dealing with in the Universe.

"Of course, Reich's work is valuable. All work that probes the nature of reality and gives some insight into it is useful. Now, we may decide, in a decade or two, that a particular description like Reich's or my own Dyadic Model, isn't comprehensive enough or is on the wrong track. Who knows at this point? I think Reich was looking at some valid phenomena. Whether he really has a powerful answer or not, I don't know. I don't find his work useful as a description that gives insight into what we are trying to understand. But I don't want to dismiss it."

The thoughts of a pioneering scientist who walked on the Moon as part of the American space program concerning the work of a pioneering scientist who died in an American prison are significant. One never knows where the investigation of the UFO phenomenon will lead or what will be learned. Dr. Mitchell concluded his remarks by sharing his views on two topics: the Mindshift Hypothesis and the Alien Agenda.

"I don't think there has been a government campaign to re-educate the public concerning extraterrestrial life. However, I think there has been a movement to do that. But I don't think it's been the government. Actually, there have been many such movements. *I'm involved with that!* I am trying to help people understand the possibility and the *likelihood* that *Life is ubiquitous.* But this is not being done as part of a coordinated or concerted effort in any way.

"Yes, there is an ET agenda. But I don't know what it is. There are people who claim to have been abducted by aliens. They claim to be reporting the ET agenda. But until we have more evidence of ET presence, I can't know that this information is credible. There is no evidence that the ET agenda is being kept from us in any systematic way.

"I think that, if I read the evidence correctly, *the clandestine group* that I have been talking about has been propagating an enormous amount of *disinformation* to hide the fact of ET presence from the people. And to construe this phenomenon in their own way for their own purposes."

Ufology is filled with fascinating sidelights to consider. For example, Dr. Edgar Mitchell's family moved to Roswell, New Mexico, when he was young. On his way to school as a young boy, Dr. Mitchell used to walk past the home of Roswell's resident "mad scientist," Dr. Robert Goddard, America's first rocket scientist. Mitchell was a seventeen-year-old living in Roswell in 1947 when an alien craft did or did not crash on the outskirts of town. At a recent lecture, Dr. Mitchell was asked if he had experienced any "missing time" related to the historic Roswell crash. He laughed lightly and insisted he had not.

## From the "Cosmic Watergate" to the "Cosmic Kindergarten"

Stanton T. Friedman holds both a B.S. and an M.S. in physics from the University of Chicago. He has worked as a nuclear physicist for General Electric, General Motors, Westinghouse, Aerojet General, and TRW. He is most definitely not, to use his own term, an "apologist ufologist." He is a forthright proponent of the reality of extraterrestrial life, alien spacecraft, crashed discs, and government cover-ups, as evidenced in his books *Top Secret/Majic* and *Crash at Corona.*

Friedman is a one of the major UFO investigators in the history of ufology. Many believe him to be the leading authority when it comes to the U.S. government cover-up of UFO information. He pulls no punches when communicating his views. In addition, Friedman has obviously mastered an impressive amount of detailed information about UFOs, which is seemingly right on the tip of his tongue. He also has a marvelous perspective and sense of humor regarding the follies and foibles of both humans and ETs.

Friedman was able to get the first witnesses to the 1947 Roswell incident to speak on the record. He has performed the most thorough investigation to date of the MJ-12 controversy. He has done important work involving the so-called "Betty Hill Star Map," a three-dimensional map that aliens allegedly showed Betty Hill on board a spaceship during her famous abduction incident. He has published scores of papers on UFOs, appeared on hundreds of television and radio shows, and delivered lectures to over one hundred professional organizations and at over six hundred colleges and universities in all fifty states and nine provinces.

Stanton Friedman is one of the few scientists involved in the investigation of UFOs and extraterrestrial life. He brings a valuable perspective to the subject.

He opened up his talk with me with disarming candor. When asked about the Mindshift Hypothesis he didn't hesitate to respond:

"There's an enormous amount of 'noise' going on in the world today about UFOs. But I think that most of what drives the work today is greed and good PR. If the government is trying to program us, they are sure doing a lousy job. My view is that the public is doing it on its own in spite of the government. *Independence Day* was not stimulated by the government. Hollywood and TV are bandwagon jumpers. I was heavily involved in the Roswell fiftieth Anniversary events in early July 1997. I noticed that the younger generation—the non-Cold War mentality, if you will—wasn't buying into the government's viewpoint. The crash test dummies timeshift story went over like a lead balloon with them.

"But that didn't stop the media. Look at the stories in *Popular Mechanics, Time, Popular Science,* and the *New York Times.* None of it was very accurate. It's a mixed bag with the media. But I saw again how viciously the Air Force attacks Roswell. They have a one-thousand-page report, *The Roswell Report: Truth vs. Fiction in the New Mexico Desert.* It was Colonel Weaver of the Air Force who supplied the fiction. And then we have Captain McAndrew's *The Roswell Report—Case Closed.* More fiction! In both cases, they get nasty, with massive misrepresentations of the facts, even about me. The Air Force is doing its damnedest. The *New York Times* has gone along with these reports, as have so many other major members of the media.

"There was a Ph.D. thesis done back in 1971 or so on the way the press covers flying saucers by Dr. Herbert Strentz of McGill.

He looked at ten thousand clippings and had some scathing remarks to make about the inadequacy of press coverage. About the failure to choose people who knew anything about the subject; about laziness. Frankly, I see little sign that this has changed.

"For over thirty years, I have probably answered more than thirty-five thousand questions about UFOs. I haven't seen any terribly substantial difference in how people respond. They are ready to listen to the facts from scientists dealing with the evidence. I don't think the 'Forces of Evil' are just in government, with people who specialize in disinformation. I think we are also stuck with some old-time media big wheels who can't admit that they have been wrong about the biggest story of the millennium! That's a major difficulty."

Friedman noted that, except for Budd Hopkins, there were not many major, authoritative ufologists in and around New York City, the media capital. In his view, this lack has a definite, negative impact. "There's a 'David Susskind Syndrome' in the media. He had a talk show years ago. I was involved in one show and gave them everything they wanted. Betty Hill and John Fuller for a good abduction case; a good witness for the Coyne helicopter case, which had occurred just a few weeks before; and copies of all my papers. They wanted a good skeptic. I told them there weren't any, but I gave them Philip Klass.

"Between segments of the taping, David Susskind said to me, 'I read the New York Times and there is nothing in there that says these things are real.' I call this the Susskind Syndrome, and many media people are subject to it. Everyone admits that if UFOs are real, it would be important. Nobody's ever argued that. But if they were important, they say, then I—fill in the name— would know about it because I take great pride in keeping up with what's important. I don't know about it, so it can't be real and anyone who says UFOs are real must be some kind of a nut."

According to Friedman, there has been a complete failure by the major media to investigate what he calls the "Cosmic Watergate." In his view, if the Washington Post put as much effort into the UFO story as it did into the Watergate investigation, the full truth of the matter could be before the American public in six months. "They could start with somebody like me. The data is there. But no, they say, 'If it were true, we would already have covered it. We haven't covered it, so it can't be true.' I've written some scathing responses to articles in the *Washington Post* and in the *New York Times* but they're stuck at the level of referring to people as 'UFO buffs' or 'conspiracy theorists.' Interestingly, when the *Times* covered the recent Air Force Roswell book, they mentioned three books about Roswell, but not mine, the only one by a scientist. The *Times* didn't interview any of the people involved; it wasn't necessary. They bought the Air Force story hook, line, and sinker."

Although he does not subscribe to the hypothesis that there has been or is now a government reeducation campaign, Friedman has no doubts about the existence of a secret government UFO group along the lines of the clandestine group discussed by Edgar Mitchell. However, he does not agree with the views expressed by Philip Corso. Friedman said, "In my book, *Top Secret/Majic,* I say that the original Operation MJ-12 documents were legitimate. I certainly think there was an MJ-12-type group. The analogy is with the Manhattan Project, an interservice operation. *I think there still is such a group.* I don't believe that Corso is the great knight on a white horse. To believe him, you'd have to believe that the MJ-12 group did nothing between 1947 and 1961, which is nonsense. Those men were *doers.* Just look at the twelve of them. I was very unsatisfied with Corso's book. I think it was an exploitation book—no index, no references. On one radio program—we were all on by phone—I asked Corso how he knew the exact date—July 6, 1947— that he saw the alien body from Roswell at Fort Riley. Was it from notes or a diary? He said, 'I know when I was transferred to Fort Riley.' Well, that was in March 1947. That's no answer! I also asked him how he knew the names of the Control Group, or MJ-12. He said, 'Well, there were about a dozen boards connected with the National Security Council.' That's no answer! I think he got the names from my book.

"As to the basic notion, did the government recover two crashed flying saucers in New Mexico, one near Corona, one out on the Plains of Saint Augustin? Yea, verily! Did the government respond by, among other things, establishing a top-secret group named Majestic 12? Yes! I'm sure that this group asked for advice on the public impact of flying saucers. You can understand their covering it up—*at the beginning.* You've got two downed alien spacecraft, a bunch of alien bodies. You know they are not

Russian or German or whatever. And you don't know what they want. You don't know where they are from. And, most importantly, how do these spacecraft work and how can we duplicate that? That *has to* be kept secret. You can understand that.

"I think that they also would have had people looking at the psychological impact. A lot of people don't seem to be aware that there was a group called the Psychological Strategy Board. It was a CIA operation in the last couple of years of the Truman Administration, about 1950 to 1952. When I first went to the Truman Library decades ago, ninety percent of their stuff was still classified. After some time, it went down to fifty percent. Now I understand it is down to twenty percent. They were really concerned about losing the Cold War to the Russians. understandably. Only a few people knew what was going on. This was kept secret through compartmentalization. They would almost certainly have asked for a psychological evaluation by psychiatrists and social scientists to the question, 'How will the public react?'"

To Friedman, propaganda was part of the game being played at the time. He thinks it is naive to think otherwise. "For anybody who thinks it wasn't, remember that we did it during World War II. In the beginning of the war, in 1942, the U.S. was getting its tail kicked. How did we get the people behind us? Propaganda. Today, a big deal is being made out of the Brookings report from the Brookings Institution. Only a tiny portion of that report looked at aliens. And it was an unclassified document. I doubt very much if anybody involved in that study had access to the classified material. You don't give out access to hot stuff! I have been to seventeen archives—Truman, Eisenhower, the Library of Congress, and others. At not one of them did I get to see special, compartmentalized information—Top Secret Majic, for example, or any other top-secret material.

"The Eisenhower Library admitted a year or two ago that they had a drawer full of this UFO stuff. Would they check it out for me? No! The Truman Library admitted that they had a half-drawer of this stuff. Some people seem to think I am on a mailing list for the distribution of classified materials. What I'm saying is that you *can't* get access to this information."

In 1995, an Executive Order was signed that was ignored by the Federal agencies that were affected by it. President Clinton signed Executive Order 12958 in April 1995. Few expected him to be reelected in 1996, so they did not pay careful attention to the order. According to Stanton Friedman, what Clinton did was very significant. "It's already having an impact. It had two very important provisions with regard to old information. One, with regard to material more than twenty-five years old, if in doubt about keeping it classified—*declassify it.* That's a reversal. The old rule was, if in doubt, keep classified. Two, everything over twenty-five years old shall *automatically be declassified* by April 2000 unless the agency that has the material takes appropriate actions to withhold it. What this means is that they are now scrambling like crazy. There's an enormous amount of classified information out there. Late in 1997, one hundred ninety million pages were declassified. Back in 1985, only five million were declassified. The Air Force, I was told, is reviewing one-quarter of a million pages a week. And they hope to make it by April 2000. All the presidential libraries are part of the National Archives Records Administration, and they have sent their classified material to the Archives for a centralized classification review.

"I am anticipating two things. One, a lot of interesting stuff will be released. Two, *somebody's going to make some mistakes.* We are talking about hundreds of people going full tilt. At one time, a few years ago, the National Archives had seventy-five people working full-time on declassification. With staff cutbacks, it dropped down to twenty-five people. With a billion pages of material, twenty-five guys won't make a dent. There has already been an impact from this on the National Security Agency (NSA). Some years ago, they put out a twenty-one-page Above Top Secret justification for withholding UFO documents. In 1980, eighty percent of the material they released was blacked out. In 1997, a new era dawned and the NSA has released documents that are only twenty percent blacked out. The Executive Order is having an impact."

Friedman pointed out that Clinton is the first President who is not a "Cold Warrior." He noted that Clinton seems to be seriously and honestly concerned about keeping information classified as secret, if only because of the enormous amount of money it costs to keep the secrets. "President Clinton stood four-square behind Secretary of Energy Hazel O'Leary when she released all that embarrassing information about radiation experiments on people. I heard her asked by Larry King what Clinton's view of her actions was, and she said the President was completely behind her. As someone who has worked under security for fourteen years, I am very hopeful that we are moving in the right direction."

On Friday, June 27, 1997, Friedman was a guest on a talk show in England on ITV when the U.S. Air Force crash-test dummy story about Roswell broke. "I was on a ninety-minute TV show called *Strange But True* which was set up as a kind of debate. It wasn't really a debate; it was mostly taped bits. It was myself, Timothy Good [author of *Above Top Secret*] and Nicholas Pope against three 'con men'—professors of physics, astronomy, and psychology. Edgar Mitchell was on with us as well. And so was Soviet General Popovich, who spoke about his UFO sighting. All during the show, viewers could call to give their response to the question 'Do you think aliens have visited Earth?' There was one number for 'yes' and another number for 'no.' One hundred thousand people called in, and ninety-two percent said 'yes'! That is a big number.

"Two years earlier, in October 1995, I took part in a debate at the Oxford University Debating Society. The proposition was something like, 'This House believes that intelligent alien life has visited Earth.' The audience got a chance to vote after hearing all the evidence from both sides. Only members can vote. The affirmative side won with sixty percent of the vote. What is interesting to me is that I did seven lectures during that trip and forty interviews. With one exception, none of the media people knew anything about the subject of UFOs. Nobody can say that the voters at Oxford were 'brainwashed' by the media at the time."

The results of the television poll and the Oxford debate seem to indicate that the public is ahead of the media regarding the UFO phenomenon. It has also been obvious for some time that the public has not accepted the government's attempts to explain away UFOs. According to Friedman, "the public is already with us. There's no need to be an apologist-ufologist. However, some of the church people have problems. Jerry Falwell and Pat Robertson, for example. In July 1997, Pat Robertson said that people who believe in UFOs should be stoned to death."

Although he doesn't agree that there is a government plan to influence public opinion about UFOs, Friedman said that there is one noticeable change in public attitude toward life elsewhere that involves the government. "Have you noticed the change involving the one thing that the government has done?" he asked. "In the past, the government's focus was on SETI—the Silly Effort To Investigate. Actually, the Search for Extraterrestrial Intelligence. Instead of SETI, NASA now has a new program, *Origins*. They aren't talking about the radiotelescope SETI search. NASA is focusing on finding planets around other stars and on Martian microbes. They may be three billion years old, but they're exciting! Pathfinder Sojourner is looking for water on Mars. And is there life on Europa? There's water there. Is there water on the south pole of the Moon? NASA is also looking at Titan—Saturn's moon—because it has an atmosphere and there could be life there, too. NASA isn't talking about radiotelescopes or listening for signals.

"What is very interesting to me—and there is even some public discussion of this—is that NASA may get a look at the alleged face on Mars. I read an article the other day acknowledging that something real is there. With the new high-resolution cameras, and so many people interested, NASA is going to try to take a look. There is one problem—you can only look down with the camera. You can't aim it. Unless you pass straight over it, you don't get a really good view. In this sense, there clearly has been a change."

Yet, it may have been the public that changed NASA. According to Friedman, the space agency was astonished at the 100 million hits on their Martian website on the Internet. Although NASA was formerly closed-mouthed about its activities, in these tight economic times, the agency has come to realize what side its bread is buttered on. Friedman noted that "NASA suddenly realized that if they want the public to support their budgets, they have to sell the public: excite people, let them know what is happening, and don't treat them like idiots. This is happening, and this is definitely something the government has done. It is probably a very good thing."

This UFO researcher is a complex man investigating a complex subject. Some of his thoughts can catch one off guard. "I want to reiterate that I am not one of those people who believes that the government should put all it knows on the table," Friedman said. "Although I am the original investigator of Roswell, I didn't sign the Roswell Initiative. Even though I have talked about the 'Cosmic Watergate,' I don't think you can go from declassifying information about what happened in 1947 to believing we should declassify everything from yesterday on back. I think that's absurd."

In addition to the mindshift at NASA, Friedman commented on another significant change he noticed. "Here's another shift. In October 1997, for the first time ever, the director of Central Intelligence announced the total black intelligence budget—$26.6 billion. This figure does not include black-budget funding for the development of advanced technology, such as Stealth. It's a rather shocking number when you stop to think about it. I don't know anybody else with a blank check for $26 billion. And the American people aren't aware of this."

The UFO phenomenon has been with us in its present form for about fifty years now. The words of a poet-friend appropriately describe the progress that we have made, "We have not traveled far." Stanton Friedman reflected on the decades he has been a UFO researcher.

"I started my research in 1958 when I read Ruppelt's book. I gave my first lecture in 1967 in someone's living room. I spoke to Congress in 1968 before the Condon report came out. Just about all of my lectures start with a standard statement, 'I am not an apologist-ufologist or a closet ufologist. I tell it like it is. My four major conclusions, after thirty-nine years of research, are:

"One, the evidence is overwhelming that Planet Earth is being visited by intelligently-controlled, extraterrestrial spacecraft. *Some* UFOs are alien. Most are not, but I don't care about them.

"Two, the subject of flying saucers represents a kind of Cosmic Watergate. Some few people within the major governments have known since July 1947 that two extraterrestrial spacecraft crashed and were recovered with ETs in New Mexico.

"Three, none of the arguments against the first two points stand up under careful scrutiny, even by such learned individuals as my University of Chicago classmate, the late Carl Sagan. Their arguments sound great until you look at the evidence, and then they collapse of their own weight.

"Four, we are dealing with the biggest story of the Millennium: visits to Planet Earth by alien spacecraft and a successful cover-up of the best data, such as the bodies and the wreckage, for fifty years.

"In my lectures, I discuss the Condon Report and five large-scale studies of UFOs. Almost no one has read any of them. In the Condon Report, there is no separate section on the unknowns. This should have been the focus of the whole report. There is a section on government investigations of UFOs, but not one word on *Project Blue Book Special Report #14*, the largest study ever done for the Air Force. This study showed that *twenty-one percent of the UFO sightings could not be explained.* These sightings were completely separate from the ones for which there was insufficient information. Ironically, I know some people who got involved with UFOs because of the Condon Report."

Just as the reader is curious about the meaning of the UFO phenomenon, so, too, the UFO investigator—whether a novice or a seasoned professional like Stanton Friedman—wonders about the significance of the effort to understand this mystery. He said, "People often ask me, 'Why are you so obsessed with flying saucers? Why do you do all this writing, lecturing, etc.?' There's an old Jewish notion—if you can do something of benefit and don't do it, that's the sin. If I can take advantage of my professional background as a scientist and my skills as a speaker to educate the public, then I should be doing it.

"My goal is to have my grandson's generation behave in such a way that we humans can qualify for admission to the 'Cosmic Kindergarten.' And that we will create a society that is suitable for intelligent life. I don't think we've done a very good job so far."

## Missing Time

Budd Hopkins is the man whose work—the books *Missing Time, Intruders,* and *Witnessed*—has made the terms "missing time," "alien abduction," and "abductee" household words. Although earlier UFO investigators used the technique of hypnotic regression to explore blocked or hidden memories of encounters with aliens, it is through his work that the general public has become familiar with this approach. In his collaborations with abductees, Hopkins works with expert psychiatrists, psychologists, and hypnotherapists.

Hopkins was an accomplished, recognized painter and sculptor long before he became involved with UFO research. His paintings and sculpture are in the permanent collections of such prestigious museums as the Whitney, Guggenheim, Museum of Modern Art, Brooklyn Museum of Art, Hirshhorn, and Carnegie-Mellon. He has written articles on painting and sculpture for major American art magazines, and also been the recipient of awards from both the Guggenheim Foundation and the National Endowment for the Arts.

His attention was first turned toward UFOs in 1964 when he had a personal sighting. His research activities did not begin until some time later, in 1975. His 1981 book *Missing Time* took the world by storm, and he was immediately recognized as a leader in UFO research. In 1989, Mr. Hopkins founded the Intruders Foundation to handle the great volume of letters and phone calls requesting information and also help.

As he has done so many times, Hopkins sat in his studio and graciously gave his time to explain his work and share his thoughts on the secretive activity of alien beings who appear to be engaged in a program involving the study and use of human beings to create alien-human hybrids. This stretch of the path will take the reader a long way from simple lights in the sky and daylight discs.

Hopkins began his interview with me by discussing a relatively recent sighting. "A Brazilian pilot reported a sighting of a cone-shaped object in November 1996. Nothing was reported about it here. As usual. Absolutely nothing. The pilot's estimate was that the object was about three hundred feet long at its base. And the apex of the thing was about two hundred feet. That's like having a twenty-story building just hovering in the sky or moving very slowly. This sighting was at eleven o'clock in the morning. It was picked up on radar; there were people watching it in the distance from a nearby beach. *Then the top of it opened and a UFO came out of it!* The pilot had a view of this for about twelve minutes. He was talking on his radio the whole time to the air base, detailing the sighting. He even got on his cell phone and called his wife to tell her about it. This pilot offered an absolutely straightforward account of a sighting of an extraordinarily large object hovering in the air, and it made all the Brazilian papers. But nothing here."

One may feel that the American media ignored this UFO report because they were not interested in a foreign story. However, in recent years, newspapers, magazines—and especially television—increasingly pick up accounts of events from around the world. But the press only chooses certain kinds of events—a raging fire, a catastrophic flood, a massive earthquake, a child who has fallen down a well. In other words, the press is only interested in "hard news." To the mass media, fires, rapes, murders, and sex scandals are the stuff the First Amendment is made on. The possibility of life from another world visiting Earth will have to wait for a slow news day. But the major media outlets do not even report what is taking place in their own backyard.

According to Budd Hopkins, "There was an extraordinary case of a large object hovering over the water in the harbor here in New York City. It was *huge*. All the water beneath it was roiling. And there were eight or nine witnesses. It was seen from the Staten Island ferry. It was enormous. Again, I know of nothing that appeared in any of the newspapers, even though eight or nine witnesses that I know of came forward to tell what they saw. I don't usually look into sightings, but I am going to look into this one because it's so strange.

"In March 1997, there was a sighting of a huge object near Phoenix, Arizona. I spoke with a man in Brazil at a UFO conference who runs a UFO reporting agency in Seattle. He has an 800 number that is used by air traffic controllers or the local police and others to alert him to sightings. He was getting calls from people in Phoenix who were tracking this huge object. He gave us an account of when the calls came in and where this thing was moving, from Nevada down to Arizona. The callers were describing what the object looked like. It was extraordinary. This story was completely buried until June 1997, when *USA Today* did an article. Some of the media picked it up after that because there were videotapes available. I think the story even made Dan Rather's broadcast on CBS."

One question keeps recurring to anyone who looks into the UFO phenomenon: what does it all mean? It is like having the pieces of the puzzle to a big picture, but no guide as to what the picture is. It is as if we do not have the proper reference points with which to frame the picture. Therefore, the overarching pattern remains elusive. As interesting as any individual piece of the puzzle may be—a strange light in the sky, a daylight sighting of a craft, an abduction consciously remembered—the mind cannot comprehend the unity of the phenomena. The implications are staggering, but the reality remains hidden.

Hopkins has given considerable thought to shifting human reactions to UFOs and life elsewhere. "What has interested me is that there is a kind of two-track situation going on. On the one hand, there is a purely intellectual acceptance of the reality of UFOs. The most recent polls indicate this. One poll showed that seventy-eight percent of Americans believe the government is lying to us about UFOs. Another poll showed that twenty-eight percent believe that we have already made contact with extraterrestrials—twenty-eight percent! I was astounded at that. I would have thought it would be about five percent; if you pushed it, maybe seven or eight percent. We have these astonishing numbers of people who, as a matter of course, accept UFOs. This has kind of seeped into the bloodstream because of TV, movies, ads, etc. That's the first track: intellectually, the belief in UFOs is there.

"The second track is, 'What does this mean to the cultures of this planet?' No one seems to want to think about what this might mean to them personally. There's something very strange about it. I was thinking how people have an incredible ability to not believe that the future is going to be any different for them. I was thinking of how many German Jews did not really think that things were going to be that bad ultimately. So they never made an effort to leave Germany. Some left, but there were many who just stayed there. There was an assumption on the part of many, many people that, as bad as it seems right now, it will be okay."

To Hopkins, human denial is evident in the rise in unsafe sexual practices among gay men, despite all the knowledge of the health risks. This denial is also evident in the behavior of teenage girls who engage in sexual intercourse without protection, all the while believing they will not get pregnant. In addition to these personal situations, he noted the big societal problems that face us, such as global warming, the death of the rainforest, and other ecological threats. Hopkins observed, "In a certain sense, all of these things, collectively, are swept aside. We can believe it, but only intellectually. We don't let it affect our lives. It is very difficult, for some reason, for people to follow through on a set of beliefs or realizations that they have.

"It seems that what is going on regarding UFOs is not in any sense going to benefit us. This is not to say that there is anything malevolent about the UFO phenomenon. But it is so *omnipotent*. And it has its own agenda that it is following. It is very difficult to think of it as something we can be casual about. No one knows what will happen. *No one does.* There is a kind of methodical quality to the UFO occupants' behavior. We know for a fact that their behavior is largely covert and deceptive. The basic techniques the UFO occupants use are manipulative and deceiving. They use screen memories or prevent people from remembering. The aliens keep people calm. The aliens say, 'Don't be alarmed. This won't hurt.' And then it hurts!

"There's a kind of manipulative deceptiveness which is different than malice. It's more akin to the tricks a dentist uses with a five-year-old to make sure he can do his operations. The dentist may use a lot of deception along the way. It certainly seems that, rather than being a series of humanistically sympathetic procedures, it's basically a way in which the procedures can be done more effectively by the aliens. The UFO occupants show *absolute indifference*. I don't see hostility. Nobody does. But there is a single-minded agenda. The single-mindedness of it is very clear. And it is carried out with great efficiency.

"Of course, there is also gigantic harm being done. The military might call it collateral damage, such as when you aim for a radar station and you blow up an orphanage. The collateral damage here is that people are taken, made as calm as possible, tranquilized, and then returned. Yet the psychological fallout, as far as I can see, is extremely damaging."

Those who write about the abduction phenomenon, either from personal experience or from working with abductees, portray two general impressions of the consequences of the interaction with aliens. On the one hand, there is *trauma* involved. Investigators such as Budd Hopkins and Dr. David Jacobs are greatly concerned about the harm done to the people they work with. On the other hand, there is also *transformation*. Investigators such as Dr. John Mack and Raymond Fowler highlight the transformative nature of the interaction with aliens. This is not a clear-cut, black-and-white dichotomy but it seems a fair description of two views of ET-human interactions.

Hopkins noted, "The experience is *always* transformative. There is no way that you can go through this untransformed. I've discussed this with John Mack. He told me, with excitement, of three people he has worked with who came to him frightened and confused. He said that after working with him for only three months, they had resigned their everyday, nondescript jobs and each one had taken a job in the environmental field. They felt the aliens wanted them to help the environment. I said to John, 'That's terrific that that happened. But if the aliens had them for, let's say, twenty-five years and they were nervous wrecks, and you worked with them for three months and they've changed their whole lives—*who gets the credit?*'

"As far as I'm concerned, the improvement is not something that came from the aliens. It came from working with John Mack. It's a *human* thing. Everyone of us wants this thing to be transformative. But it's not transformative *per se*. It is only when a helping hand is offered—whether it be John Mack or David Jacobs or myself or any of a number of people—that we can help them remember, help them build ego strength, and get them the support group they need. That's when the *healing* takes place. I think that the aliens have been systematically and inadvertently damaging lives. The transformative experience we all want has nothing to do with the aliens whatsoever. They are actually responsible for enormous amounts of damage."

Although the abduction phenomenon has made its way into movies, books, magazine articles, and television talk shows and dramas, it is only recently that UFO investigators have been looking into the subject. No one actually knows when the abductions began or how long they will continue. Neither does anyone know with certainty what the alien abduction program is about. Budd Hopkins expressed his frustrations with this lack of knowledge.

"We are always tangled up because we didn't even know the abductions were going on in any systematic way until the Betty and Barney Hill case. And this case is almost twenty years into the so-called modern era. For all that time, no one even thought about abductions. We were trying to collect sightings. That's like trying to get the license plate on the getaway car without having figured out what the crime was. Once we began to study abductions, we still didn't know what to ask, how to approach them. It is only slowly that we have been able to begin to understand the levels of this. We don't know whether the abduction phenomenon has changed organically or whether what we know is only what we have been able to find out. My suspicion is that things haven't changed much. Except perhaps in volume. My feeling is that the same damn thing was probably going on in 1931 as today. Except perhaps in greater numbers now."

Although he sees increased public acceptance of the UFO phenomenon, Hopkins is uncertain about the Mindshift Hypothesis. "It's certainly a possibility. But I don't see any evidence for it. Historically, we have received nothing from official sources but mixed signals. Can you say they are educating us when *Independence Day* and most of the other movies are monster movies? They are going to come and blow us up—is that what the government wants us to know? Or people have said to me, 'Why did the government let CBS make *Intruders*?' CBS figured out this was a hot topic. It was for money! Is the government letting Fox do *The X-Files*? That show says the government is lying to you, as did *Intruders*. These were money-making propositions.

"Today, you can hear anything, especially on the Internet. You don't have absolute control out there. The Internet is essentially like a newspaper with eighty-five thousand columnists and two reporters. You can hear anything. But I have seen no evidence that there is any kind of systematic release of information.

"For example, Corso's book is a total, one-hundred-percent fraud. I wrote a review of it. I ask myself, was there hardware captured from the aliens? I believe there was. I believe that something crashed at Roswell. Next, I ask what would have happened then? Well, the hardware would have been reverse-engineered. Then another question comes floating in. What was Corso's role in 1961, fourteen years after the crash? The top-level scientists would have pounced on this material and would not have let it go. These would have been the most important objects on Earth. The message of Corso's book is that nobody did anything until he came along. That is foolishness. I think that probably by 1961 everything was so institutionalized that Corso would have just been a tiny cog. In fact, I think that if the book were true, he probably wouldn't have been able to write it. But I don't think the book is disinformation."

There *is* a change going on, according to Hopkins. But he sees it as arising, not from the behind-the-scenes activities such as described in *The Day After Roswell*, but from the hard work of many individuals over the years. "The change that is taking place is due to the fact that so many of us have been presenting evidence on television—our biggest audience—and through books, radio, and so forth. On television, people get a chance to look at faces. This is what TV does best. You decide whether somebody is lying to you or whether that person looks like he's telling you the truth. That's essentially what happens when you present an abductee who tells his or her account.

"I think it's the evidence itself—and not any kind of government activity—that is changing things. If anything, the government only attempts to ridicule the evidence through such efforts as the recent crash-test dummies story. Which has backfired. I think the government's attempts to keep the lid on are not only poorly done, but also, it's a Pandora's Box. You can't shut the door again once it's open. I think it's the material itself that has changed things. A lot of people have resolutely presented this material effectively. It's due to Jerry Clark, Stan Friedman, David Jacobs, John Mack, Ray Fowler, and many other people who have presented this material effectively in conjunction with the evidence and testimony.

"Today, people stop and think, 'Now, wait a minute. As our space program goes on, it makes perfect sense, it's more logical that this UFO phenomenon is going on.' And the government's attempts to cover this up have been ludicrous."

To many people, it is inconceivable that there can be efforts behind the scenes to accomplish broad, long-range goals. Military people meet daily to plan strategy for the coming decades; corporate leaders develop five-year business plans; political groups hold conclaves to plot the next election campaign. These activities may result in success or failure, but they can usually be kept secret when so desired. Yet, when it comes to UFOs, it is difficult to imagine humans working together, out of the public eye, to achieve a desired end. The words "paranoid" and "conspiracy" are often used to describe such hypotheses as a re-education campaign concerning UFOs, even by pioneers in the field.

To Hopkins, "The trouble with any kind of paranoid view is that paranoids always wind up granting the opposition almost total intelligence and cleverness and omnipotence. *They* can figure it all out and know how to do everything. *They* are unbelievably and uncannily perfect. Yet, in this case, the cover-up has been poorly managed, ineffective, and the evidence presented by the other side has gotten through. I did a program in Canada not too long ago. Viewers could call two numbers to respond to the question 'Do you believe we have made contact with aliens or we haven't?' Granted, it's a self-selected process, and it may not mean anything. But ninety-two or ninety-four percent of the callers believed we had already made contact with aliens. I was astonished. You are bound to get a majority because skeptics aren't going to watch the program. But still, that was a significant number."

As researchers such as Budd Hopkins continue with their investigations, those numbers will surely increase. Through his efforts, and those of his colleagues around the world, humanity will gradually come closer to an understanding of the cosmic phenomenon that has entered its earthly existence.

## The Doctor of Space

On May 4, 1995, the *New York Times* ran a piece in its national section under the headline, "Harvard Investigates a Professor Who Wrote of Space Aliens." True to its bias on the subject, the newspaper of record opened with a lead paragraph that read, "In a rather bizarre example of peer review, a committee at the Harvard Medical School has examined the work of a tenured professor who wrote a best-selling book about people who say they were abducted by diminutive, large-eyed, gray-colored creatures from outer space and forced to have sex with them."

The tenured professor in question is John E. Mack, M.D., a professor of psychiatry at The Cambridge Hospital, Harvard Medical School, and founding director of the Center for Psychology and Social Change. He is also the founder of the Program for Extraordinary Experience Research (PEER), an organization devoted to the study of the abduction phenomenon and other anomalous experiences. Dr. Mack's bestselling book is *Abduction—Human Encounters With Aliens*. In 1977, Dr. Mack won the Pulitzer prize for his biography of T. E. Lawrence, *A Prince of Our Disorder*.

Dr. Mack's book *Abduction* is dedicated to "Budd Hopkins, who led the way." In the fall of 1989, a colleague asked Dr. Mack if he wished to meet with Budd Hopkins, a New York artist who worked with people who believed they had been abducted by alien beings and taken aboard spaceships. At first Dr. Mack declined, believing the whole matter to be crazy. But on January 10, 1990, he met Budd Hopkins.

It is now Dr. Mack's opinion that we live in a universe—or universes—in which there are many intelligences from which we are cut off. We have lost the ability to sense them. Dr. Mack's expert diagnosis of his patients is that *they are telling the truth.* Despite the seeming impossibility of many of the things the abductees—or "experiencers," to use Dr. Mack's term—report, what they have to say cannot be attributed to psychological problems or mental illness. These men and women are not hallucinating or dreaming, although many people fervently wish they were.

To Dr. Mack, the testimony of the experiencers has the potential to transform our view of who we are and of our place in the Universe as profoundly as the discovery of Copernicus that the Earth is not the center of the Universe. Dr. Mack's view of the whole UFO abduction phenomenon is evolving and shifting as he grapples with new information.

"I have increasingly come to see the UFO phenomenon in much more complex and spiritual terms than I did originally," he says. "My work has probably gone in a direction that will increasingly irritate my 'nuts and bolts' colleagues. I tend to think of this phenomenon more and more in evolutionary, ontologically complex terms. I don't think so much any more in concrete terms regarding the aliens. I don't see the experience as if it is totally material. The experiencers, however, feel that way. They feel as if they were taken away bodily. They have seen hybrid babies. It really happened to them; it's not a fantasy or a dream.

"But, to me, that doesn't mean it happened altogether in the material world. Nobody has actually ever literally seen hybrid beings on Earth. Or seen photographs of them. They seem to exist in some form or another, and the experiencers find them very real. It seems real to the people who have had the babies—the so-called babies—and seen them on the ships. For example, a mother may get very upset when she has been told by the aliens, or when she feels, that the hybrid is her biological offspring and then she cannot get back to it and has no control over the situation. It seems very real. Yet, at the same time, we have no concrete physical evidence for it.

"Linda Moulton Howe has called her latest book *Glimpses of Other Realities*. The UFO phenomenon gives us *entree* into other dimensions or other domains of being or reality. I have increasingly come to see that this whole thing may be represented concretely in images and experiences, but that doesn't mean that it is material."

Dr. Mack's views are those of a classically trained psychiatrist with thirty-five years of clinical experience. On top of that, he has worked with hundreds of abductees or experiencers since he became involved with this subject in 1990. His reflections represent his emotional and intellectual reactions to startling, challenging information. His general evolution toward a spiritual point of view is not unique in ufology. In different forms, this viewpoint is shared by other UFO researchers. But what do the experiencers think and feel?

Dr. Mack says that "The experiencers say it is all very vivid and material. They wouldn't agree with what I am saying. When I share my thoughts, they might say to me, 'Oh, that's interesting. . . . But I thought I was really on the ship.' However, some people are not actually missing while they are having their experiences. It is a psychological experience. But it's peculiar because in other cases the people are actually *reported* missing.

"The UFO phenomenon penetrates variably into our material reality. It can cross over into it literally, or it can be present only experientially. It is very hard with the Western mind—or at least this Western mind—to find a way of *languaging* this. That's why I stay away from the media now. They want to know 'Was this real? Or was it not real?' All the ontological subtleties are not something the media know how to communicate. They are not sound-bites."

The proper role of the media regarding UFOs is difficult to assess. It seems that either the UFO phenomenon is ignored or it is misrepresented when it is covered. There is a sense among most researchers that it is important for the public to be aware of this information. But there are serious obstacles in the way of communicating about this mystery.

Dr. Mack observed that "We don't even know what we want people to be aware of. Do we want people to be aware that there is in fact a phenomenon that is in some way real? I don't know what happens in the media, but there is something I call 'Spielbergism.' You take something that is real. Something about which there is a great deal of conscious interest—and anxiety—in the public. Something that is powerfully real and important. And then you fictionalize it, sensationalize it, trivialize it, sentimentalize it, ridicule it, exaggerate it, and over-concretize it. You make it entertainment in one way or another. It can be good entertainment or not well done. But it still turns the reality into a fantasy. Whether that tends to advance or to slow down the appreciation of the reality is unknown. Did H. G. Wells's futuristic novels increase knowledge of the possibilities of time warps and time travel? I don't know. It's not that simple."

The role of the media is related to the hypothesis of the re-education campaign concerning extraterrestrial life. Does the media act independently, motivated only by money? Is the media used by people with a clever plan acting covertly? Or does the truth lie somewhere in between? Is there even a campaign at all?

Dr. Mack is skeptical. "I don't think there has been a campaign. On the contrary, I think there is a total lack of rational education of the public regarding this reality. If you follow the flow of information on this subject in the media, you will get a range of material from earnest coverage of the facts to complete dismissal and ridicule. There is no evidence that there is any sort of cohesive education. The government is totally silent or makes up false stories that are not plausible, such as about Roswell.

"Some people think that there is even the opposite of an education campaign—*a systematic disinformation campaign* to confuse the public about the reality of UFOs. It's like the old joke, 'What do you think of Western civilization?' And the answer is, 'I think it would be a very good idea.' I think it would be a very good idea for responsible people—a combination of clinicians, scientists, government people, and media people—to get together and present some sort of modest, thorough overview of what's going on in this field. We don't know the realities altogether. But there is a lot of information that could be communicated to the public without a lot of drama. This could go a long way to reducing fear. For example, there is a lot of evidence that there have been encounters. There is no evidence or suggestion—except for David Jacob's new book *The Threat* which postulates a take-over kind of thing—that these beings or whatever this phenomenon is meant harm to the human race. If anything, my work suggests that there is some *awkward effort to preserve life going on.*"

According to Eastern thought, the farther one travels, the less one knows. The further one delves into the world of the UFO and extraterrestrials, the less certain one becomes of one's intellectual and emotional footing. Is it real? Did people see physical spaceships? Were they taken bodily into alien craft and given hybrid babies to hold? If it is not real, why is there a cover-up? Is a disinformation campaign significantly different from an informational campaign? How easy it is to fall through the looking glass.

To Dr. Mack, it is "a mistake to literalize this too much. The problem with this whole field is that the deeper I get into it, the less literal I become about it. Which isn't to say that the experiences aren't literally real. For example, there's the whole business of the demography of aliens. People have experiences with different types of beings, and they do have some properties that differ. Some people have experiences with several different types working on the same ships all at once—luminous ones, gray ones, tall beings, praying-mantis types, reptilian ones, robotic ones. They seem to be able to co-exist with one another. But I think you can be too literal with this, such as by thinking that there are some flying saucers with grey beings and others with luminous ones. The literalizing of this along left-brain categorizations—I don't think it works like that. I know there is a whole literature of the various alien types. Some say there are about thirty civilizations. I would have to be careful about lending them the same material status we give to the indigenous groups in Africa, for example.

"There seems to be some sort of evolutionary element here. The word I seem to come back to is evolution. Whether it is evolution of consciousness or evolution of biological forms or of Spirit, I'm not sure. But there does seem to be some sort of *process* of possibly biological evolution. Although here, too, I don't know how literally to take this. There could be the evolution, not literally, of a hybrid race that could survive the ecological holocaust that is underway here. But it seems to be more the *idea, the metaphorically embodied idea of evolution,* that this all seems to be about. It's almost as if we were having a course in the evolution of species. It represents the future should we not choose a viable future. It is what the future will look like if our future is not viable. Of course, what we're now doing is not viable."

We know that in the past species have evolved that were unable to survive in the world. They were unable to create a home for themselves. They were biological "dead ends." We also know that some species—such as the dinosaurs—survived for hundreds of millions of years, only to perish because of dramatic changes in their environment. We humans are now responsible for the death of untold numbers of species. Will we also be responsible for our own demise? Does the UFO phenomenon offer any insights here? It may.

In his work, Dr. Mack has discovered that "experiencers see images of vast destruction. The destruction of beauty and living systems. They feel a great sense of horror at the apocalyptic disaster that is being shown to them by the aliens. But this is only a metaphor or a concretization of what is happening on our planet anyway. There is a controversial spiritual opening that is associated with these experiences. The people involved feel connected beyond themselves with these beings and open to a sense of connectedness with Nature. It is all part of this evolution of awareness and consciousness. It seems to be intrinsic to the UFO phenomenon. This is not to say that it may not also be traumatic. But this evolutionary element seems to be part of it.

"Why this is happening is awfully hard to answer. I tend to look at it contextually. In the context of the systematic destruction of the Earth's living systems, by this one species, the UFO phenomenon seems to be the Universe's awkward way—or at least this region of the Universe's response—to our situation. So much of the material I get from my clients is related to this ecological crisis. When you put that together with the *fact of the ecological crisis,* it makes sense.

"Think of this connectedness as existing not just at the cellular level or at the personal level, but at *the cosmic level.* This one rogue species, namely us, has gotten out of balance and is treating Earth as if it were its own property: carving it up and making housing developments, destroying thousands of square miles by mining for minerals, burning all the forests, polluting all the oceans—taking this sacred jewel in the cosmic crown and treating it like a piece of real estate and destroying it. If we could truly experience that there is this sacred connectedness in Nature that extends beyond the Earth itself, then it would make sense that our activity would elicit some kind of response from the Universe, some feedback of some sort. I think of the abduction phenomenon in that way, as a kind of cosmic feedback."

The thought of life elsewhere is just barely becoming acceptable to large numbers of people on Earth. Fossils of ancient microbial life on a ten-thousand-year-old Martian meteorite are not threatening to most people. The possibility of life on the moons of Jupiter, or on unseen planets revolving around distant stars, does not fill the average person with alarm today. But the idea that advanced intelligences may be visiting our planet for reasons known only to them is another matter. The concept that these visits may be in response to our poor stewardship of Planet Earth is something far beyond the implications of extraterrestrial reality itself. How can we begin to understand what is happening?

Dr. Mack said, "I would like to invite people to move beyond a literal, material world view. To consider that some phenomena can manifest in the material world—like a UFO—and, at the same time, *not be of the world.* It can be what I call a crossover phenomenon: something that seems to emerge from another dimension and just shows up here. I'd hope that there could be an appreciation of this crossover phenomenon. It happens with near-death experiences, when the psyche seems to project things psychokinetically. I hope that the professional community—and the scientific community in particular—would open up to these ontological ambiguities.

"I think this phenomenon can serve to open us up once again to a sense of our proper place in the Universe, a sacred place. It opens consciousness to the spiritual realities around us. We are less likely from that place to treat the Earth simply like private property, as a resource to be mined or excavated or taken over. If the Earth can once again be experienced as *sacred*—which is what I tend to think happens with the experiencers I work with— Nature again becomes sacred. Then a more shepherding, steward-like relationship to the physical world can occur. I think there is a potential that this phenomenon holds out for a greater sense of responsibility. I see the UFO phenomenon as a very powerful tool for consciousness. That's my main interest at this point."

The human animal is cut off from its roots in Nature. This has led in turn to the development of an irrational philosophy— expressed popularly as *Man against Nature*—which is being put into practice today, causing the ruination of Earth's ecosystems and the extinction of living species at a phenomenal rate. If the UFO phenomenon—even inadvertently—helps the human animal reconnect with Nature within and without, it will prove to be a gift from the Universe without compare.

## From "Nuts and Bolts" to the "Paraphysical"

Raymond Fowler may have entered the Twilight Zone of ufology. Just when the UFO and abduction phenomena have moved from the supermarket tabloids to the *New York Times*, *Washington Post*, and the *Wall Street Journal*, Fowler has written a number of books detailing human-alien interactions with a very high degree of strangeness.

Fowler has been a major UFO investigator for over thirty years. He is respected the world over for his unbiased reporting and dedication to the facts, wherever they may lead. Among his many books on UFOs are *UFOs: Interplanetary Visitors*, *Casebook of a UFO Investigator*, *The Melchizedek Connection*, *The Andreasson Affair*, *The Andreasson Affair—Phase Two*, *The Watchers*, *The Watchers II*, *The Allagash Abductions*, and most recently, *The Andreasson Legacy*.

In addition to his research and writing about UFOs and extraterrestrials, Fowler operates the Woodside Planetarium and Observatory in Wendham, Massachusetts, where he conducts public classes on astronomy and UFOs.

Over the course of his work in ufology, Fowler has seen the phenomenon become simultaneously more accepted and more bizarre. At the beginning of the "modern era," any mention of the topic of flying saucers brought derisive remarks about seeing "little green men." Today, people everywhere have a more open mind about the possibility of intelligent life from elsewhere visiting Planet Earth.

Following the facts as they emerge in his research, Fowler has linked the UFO phenomenon with such controversial topics as out-of-body experiences (OBE), near-death experiences (NDE), and other anomalous experiences. According to Dr. Barry H. Downing, author of *The Bible and Flying Saucers*, "No one has done more than Raymond Fowler to explore the religious dimension of UFO abductions in a scientific way." And Kenneth Ring, Ph.D., author of *The Omega Project* and a pioneer in the study of the possible relationship between UFOs and the near-death experience, believes that "Raymond Fowler . . . may well have deciphered the ultimate nature and meaning of the baffling but irresistibly intriguing UFO abduction phenomenon."

In Fowler's book, *The Watchers*, the grey aliens who interact with Betty Andreasson tell her that they are "the gardeners" who watch over "the garden"; that is, Planet Earth. It is interesting to note that Fowler, a man who has contributed so much to an understanding of the UFO phenomenon, is himself an avid gardener.

Fowler began his interview with me by sharing his reflections on the role of the media. "In the last several years, there has been a raft of books written on UFOs, OBEs, NDEs, and other related subjects. This indicates a greater interest in these phenomena today. The books would not have been published if there were not a market. I think basically the media are responding to this interest. There also seems to be an *Internet explosion* concerning such information. The Internet is playing a huge part in our changing views of our place in the Universe. Researchers who were separated by many miles and in different countries, who have been doing independent research, now have contact with other researchers and can share their resources and ideas over the Internet. I think the Internet will continue to be a great aid in researching UFOs and other phenomena as well.

"Of course, this opens the door for a lot of material that shouldn't be on the Internet, or in books as well. One has to look carefully at the credibility of the researcher and the credibility of the participants in the various reports. However, there is an information explosion going on now and, frankly, the major media have caught on to this. They now seem to have a greater interest in the subject. From the last great UFO wave in 1973 until the present, the major newspapers have ignored the UFO phenomenon. Local papers all over the country were still reporting sightings in their locales. But the major papers did not. Today, even major papers such as the *Wall Street Journal* are carrying UFO stories. And so are very reputable magazines such as *Saturday Review*. Even abduction stories are carried now. Recently, the *Washington Post* ran a long article about a government lawyer who decided to come forth and give his [abduction] story.

"I think there has been a paradigm shift, or whatever you want to call it, especially in areas where we as a society have been taught, by the scientific community and the theological community, that certain things did not exist. But now, millions of people are coming forth with UFO stories, NDE stories, experiences with angels. Angels in particular are a big hit today. All of these things, which we had shifted outside of reality, have come to the forefront suddenly because of peer pressure. People are saying, 'I don't care what the scientific community is saying. I don't care what the theological community is saying. There are just too many of these things happening to too many people.' This is having a sort of domino effect. As information gets out into the open, people who have had these experiences, thinking they were the only ones, now feel that they have a platform and that they can talk openly about their experiences."

This shift has become evident outside of newspapers, magazines, television shows, and talk radio. It has begun to make its way into "the real world" in which we live and work day after day. For example, Fowler pointed out that "Even in hospitals now, they have support groups for people who have had near-death experiences. And, in the area of psychology, there are now support groups for abductees. Also, we had a symposium on UFOs in 1992 at MIT. That would have been unheard of in years past. I think we are on the way to opening this whole area up for public consumption and public discussion. I hope that the scientific community—other than those already associated with UFO research organizations—will get the message and look into the UFO phenomenon themselves.

"But as with anything—the Internet, TV, radio—you'll get all sides on this issue. It's the same in religion, politics, or any subject. You have to learn to separate the wheat from the chaff, the signal from the noise, or however you want to put it. At least these subjects are now out there. They are part of our reality but have been ignored because of scientific and theological peer pressure in the past."

Fowler was not always on the forefront of the spiritual dimension of the UFO phenomenon. Although he is undoubtedly a pioneer in the field, he underwent his own mindshift and transformation over the past three decades. UFOs and intelligent extraterrestrial entities do not have an impact only on those humans who see a UFO or who have an encounter with aliens. Investigators are changed over the course of their investigations. Readers may even be changed by what they learn about the subject from the many books and articles that are available today. What was it like for Raymond Fowler in the early days?

"As a UFO researcher, I started out basically interested in the 'nuts-and-bolts' aspects of the phenomenon. I thought that anything outside of this nuts-and-bolts approach was outside of the realm of true research. Back in the 1980s, I would come across people who had had close encounters with UFOs. Then they would begin to talk about other things—psychic phenomena and things like that—that they had experienced. Immediately, I felt that this would compromise the UFO report, the UFO experience. I would *exclude them* from my data base because of that. I also felt that if I were to mention these things even within my own family—including some things that were happening to me—I would compromise myself as an objective UFO investigator.

"As time went on, I saw that Jacques Vallee had moved from the hardware, nuts-and-bolts approach. I talked to Dr. Hynek about this, and we both thought that something was wrong with him. Yet, as I got into the Andreasson affair, and became engrossed in that work, I talked to Budd Hopkins and others who were involved with the abduction work. I think that Budd Hopkins is still kind of cold on this part of the UFO phenomenon. I began to see that we were looking at all these other phenomena as separate entities. But they all seemed to be connected. I began to realize that it wouldn't be honest for me to continue in my research unless I included the whole ball of wax. I was claiming to be objective. And people thought of me as objective. But I really wasn't. I was rejecting the paranormal aspects of the UFO phenomenon. This realization changed me. I'm now able to look at these things. It is interesting that some of the new people who are coming on board—such as John Mack—immediately embrace that part of the phenomenon.

"There has been a shift in my own perception of what is going on, even outside of the abduction phenomenon. I came across many cases of people who had close encounters of the first kind—a UFO sighting—who immediately began to experience paranormal phenomena such as poltergeists, apparitions, out-of-body experiences [OBEs], and the like. This confused me. I could not understand what was going on."

Fowler's in-depth work with Betty Andreasson is one of the most remarkable bodies of work in all of UFO research. The intertwining themes of mystery and revelation, followed by a deepening of the mystery and a more profound revelation, make for riveting reading. This involved investigation is at the core of Fowler's thinking. He said, "When I was investigating the Andreasson Affair, Betty and her husband Bob had a shared *OBE abduction* during which they each met the same types of entities, namely the Greys, who were working with the Nordics, the human-looking aliens. There are also other accounts of OBE abductions. They seem similar to physical abductions. It appears we are dealing with something not physical but *paraphysical*.

"I then looked at my father's out-of-body experiences. I looked into Kenneth Ring's investigations of the NDE phenomenon. Ring identified some cases as 'mixed motif,' because these people had near-death experiences that were *identical* with UFO abduction experiences. To me, this whole thing is connected. It's as if we are dealing with *metaphenomena*. We are dealing with one thing but, until recently, we have been treating all the different facets of it as separate things. Without our realizing it, the whole thing—UFOs, OBEs, NDEs—may all be intimately connected.

"Many years ago, Wilbert Smith, a project engineer who headed a Canadian government UFO research effort called Project Magnet, was involved in the area devoted to magnetics. He was in Washington, D.C., and he inquired into what was going on with UFOs within the U.S. government. He was told that Vannevar Bush was heading a small group that was looking into UFOs and that the whole subject was classified higher than the hydrogen bomb. He also learned that their *initial* conclusion was that UFOs were a 'mental' phenomenon. In those days, 'mental' would equal 'psychic' today. I came across people that the U.S. Air Force was investigating for psychic capabilities. One person I looked into in detail had been tested by the Air Force for *psychokinetic abilities* and had been attending NASA meetings in Boston. I learned from this woman that UFOs were considered to be a part of a psychic or paranomal phenomenon."

One can understand why Fowler was reluctant to enter such uncharted territory. After all, when one works at the frontiers of new knowledge and begins to find some acceptance, it is extremely difficult to enter even more controversial areas of inquiry. It is natural to want one's work to be understood, one's research to be accepted. What does an investigator do when the facts being uncovered lead even further away from the accepted wisdom of the day? Fowler followed the facts.

"If we take OBEs at full face value, and accept that they really exist as people describe them, then we have reached the dimension the 'OBEer' enters, another dimension outside what we call the physical dimension. In this unknown dimension, the person having the OBE can recognize the physical but cannot interface with the physical, and vice versa. Another person cannot sense or see the person having the OBE.

"If we take NDEs at full face value, we find another dimension outside of what we call the physical. In the NDE, the essence, soul, spirit, or whatever you wish to call it, leaves this space-time dimension and goes further, down a tunnel, and enters *a world of light*. There he or she meets robed entities who, in many cases, look exactly like the so-called Nordics seen in UFO abductions. Bob and Betty Andreasson had an OBE abduction in which they exchanged their bodies for bodies of light. They were in another dimension with these robed entities and their Grey workers. It starts to all fit together.

"It seems that the physical abductions are at one end of the spectrum. The physical abduction seem to involve something like we would do if we wanted to save a species from extinction. I don't even like to call these experiences abductions. I prefer to call them operations, or something like that. The NDEs are at the other end of the spectrum. This phenomenon seems to involve where we are going next, in the next stage of our evolution. I think it is very interesting that the entities themselves, the Nordics, look almost exactly like Homo sapiens. They usually only interface with human beings at the OBE level. The Greys, on the other hand, seem to be able to go back and forth, from here to there, without any problems. The Greys told Betty Andreasson that they are a form of Homo sapiens. They are mature human fetuses, grown outside the womb, and tailored to do their work. They certainly look fetal!"

The more one follows Fowler along his path, the stranger the journey becomes. His research is uncovering material that is so exotic it makes flying saucers appear completely within the realm of the ordinary. According to Fowler, there are reports of Reptilian and Insectoid alien beings that are also humanoid in form. Fowler observed that the humanoid alien beings represent variations of lifeforms that exist here on Earth now—insects, reptiles, and mammals. "It looks like someone or something is able to use these species to create humanoid workers. There seems to be a definite relationship here. I speculate—*and I stress that this is highly speculative*—that when we die, and we have an NDE and enter the world of light, we may indeed become what the Nordics are. We may still be humanoid. We still live on. The human form may be the larval stage of what we are to become. I never thought in my wildest dreams that I would ever, ever think like this. It is by studying this material—the OBEs, the NDEs, the abductions—that I have arrived here."

Not all ufologists go along with Raymond Fowler on his journey from nuts-and-bolts research, through human abductions by aliens, to OBEs, NDEs, and life after death. Yet, he is a credible researcher who deserves a hearing. Many who once considered his reports on UFO sightings to be preposterous are now giving UFO reality a second thought. Could the same occur concerning even his most wild speculations? Are some aliens a form of Homo sapiens? Is that why they are here?

According to Fowler, "The entities claim that there is a symbiotic relationship between us and them. They say that what we are doing ecologically is going to affect them. In my book *The Watchers II*, I use the analogy of dragonflies going back to the pond and telling the larvae that if they don't clean up the pond, there won't be any more dragonflies. Because of what we are doing, they are taking ova and they are taking sperm. Perhaps they want to have a huge sperm-and-egg bank. If you accept the abduction reports at full face value, the entities seem to be creating hybrid forms.

"In my most recent book, *The Andreasson Legacy*, from the calls and letters I get, I chose at random to investigate the story of a woman who had no interest in UFOs at all. She had not read any books on the subject. But some strange things were happening to her, and she confided in her sister. Her husband wouldn't listen to her. The sister said, 'I saw something on TV like this' and took her to the library, where they found one of my books.

They got my address and called, asking for help. I found childhood UFO experiences, ova removed, interaction with hybrid children. It all came out. I felt bad for the woman. She had volunteered, but she hadn't the faintest idea of what could come out under hypnosis."

How prevalent is the abduction phenomenon? Polls reveal some startling numbers, and skeptics use these figures to make fun of the very idea. But is the basic concept of an alien program to collect and save human ova and sperm so foreign and untenable? We ourselves are now doing such things with our own technology. Is an alien hybrid program so unthinkable? Who knows what lurks in our future when the "brave new biologists" really get going and they "send in the clones." Human cloning is now possible. What could intelligent beings with highly advanced technology be capable of?

The Mindshift Hypothesis may help here. Perhaps the existence of intelligently controlled UFOs is not what the powers-that-be feared would cause panic on Earth. Perhaps it is not even the reality of superior beings visiting Earth that is the most threatening aspect of the whole phenomenon. Could it be that responsible leaders know *why* the ETs are here? Is it that information that is so alarming that it can only be gradually released to the public over generations?

Dr. Edgar Mitchell is convinced that there is a clandestine group controlling all access to the UFO information that proves extraterrestrial life exists. But he sees no evidence for an organized campaign to re-educate the public. Stanton Friedman broke open the Roswell cover-up, is convinced that the ultrasecret UFO group MJ-12 was a reality, and believes such a group operates today. But he sees no evidence of a government educational campaign. Although Budd Hopkins admits such a campaign is a possibility, he sees no evidence for it. And John Mack likewise sees no evidence of such a cohesive education campaign. Interestingly, he raises the possibility that there could be a government *disinformation campaign.* And Philip Corso asserts flat-out there has been and still is a government effort to simultaneously disclose and cover-up UFO-related information. Where does Raymond Fowler stand on this issue?

"I can see why people wonder if there is such an educational campaign going on out there. I often wonder if someone flicks a switch and says it's okay to do programs about UFOs. The media seems at times to be completely turned off. Then all of a sudden—bingo! UFOs are everywhere. And it happens almost overnight. When you look at our regular news, we just get these same capsule stories. Then you look at BBC and you see news about all these tremendously important things that are happening in the world. You get the feeling that we just get selected news.

"I wonder also if there is encouragement from behind the scenes for UFO material to get out for a conditioning purpose. I remember when the movie *Close Encounters of the Third Kind* came out. There were rumors then that the Air Force or the government—whoever *they* are—cooperated with Spielberg and promoted the idea that the ETs are friendly. I think it's a possibility that something like a campaign to re-educate the public is going on."

Whether or not there is an organized campaign to re-educate the public concerning UFOs, there is certainly a determined effort being made to do that by scores of people. Fowler said, "Each one of the people who studies the phenomenon has his own view of what's going on. For example, Stanton Freedman is certainly interested in extraterrestrial lifeforms from another physical planet visiting Earth. Budd Hopkins is interested in the genetic engineering that's going on. It's anathema for him to think the paranormal part of this is real. I am trying to look at the whole phenomenon, and I'm sure that I am not completely objective either. With *The Watchers*, I sent up a trial balloon to see what would happen. In *The Watchers II*, I went a little further. In *The Andreasson Legacy*, I looked back on the whole twenty-year investigation."

Fowler believes that humanity is facing very serious problems today and in the years to come. For example, he has reported that the aliens told Betty Andreasson that mankind and other lifeforms on Earth are slowly going to become sterile. There have been many stories recently in the mainstream press about human sperm count dropping statistically. These news stories do not prove that what aliens allegedly communicated to Betty Andreasson is true. However, it is noteworthy that she knew this information long before there was any evidence of a significant drop in human sperm count.

In assessing mankind's situation, Fowler said, "I think humankind is in a state of denial. Every week you read about the extinction of another species. All these ecological disasters keep happening, and they get on the news. People see this and say, 'Oh that's too bad . . . global warming.' And then they go on with their lives. It's almost like someone told them they had an incurable disease. They go through various stages of denial until, finally, acceptance. We are in the first stages of hearing about these things and are ignoring them because of the Almighty Dollar. If you accept the entities message, I guess you might call it, we may destroy life as we know it.

"There are those who feel that the hybrids the aliens are making are going to take over. I think that is pure speculation. In fact, *everything that I have said is sheer speculation.* Everything that Hopkins and Friedman and Jacobs and Hynek—all of us—have said is sheer speculation based on anecdotal data. We are all trying to figure out what's going on. I think it is valuable for the Friedmans and the Hopkinses and the Jacobs and the Hyneks to put out for peer review what they think about the UFO phenomenon. I hope a more coherent picture will come out of it. But I think that what researchers have to do is to be more openminded and realize that this phenomenon might incorporate a lot more than any individual's pet ideas."

## The Threat

David M. Jacobs, Ph.D., associate professor of history at Temple University, is a leading academic authority on UFOs and alien abductions. He is the author of *The UFO Controversy in America*, which remains perhaps the best history of the phenomenon since it was published nearly a quarter of a century ago. Dr. Jacobs is also the author of *Secret Life: Firsthand Accounts of UFO Abductions*, published in 1992, a book that set a new standard to which all subsequent writings on the abduction phenomenon would be held.

Dr. Jacobs uncovered important patterns in the alien abduction scenario, patterns that revealed alien mind-control techniques, physical examinations of human beings, and an intense focus on human sexuality on the part of the extraterrestrials. Dr. Jacobs personally researched every account in *Secret Life*, in which he detailed a common abduction scenario that involved physical, mental, and reproductive components. The alien operations were further broken down into primary, secondary, and ancillary procedures that entailed implants, tissue samples, "mindscans," sexual arousal, egg and sperm collection, embryo-implanting and fetal extraction, hybrids, imaging and testing, surgery, and knowledge and information transfer.

As unsettling as his past work has been to many, his new book, *The Threat*, has caused a stir in the UFO community because of its shocking conclusions regarding the ultimate purpose of the alien abduction program. The ideas presented by Dr. Jacobs may be controversial, but they provoke open and honest debate.

Dr. Jacobs started his interview by detailing the beginning of his work. "I started out in the mid-1960s. I began to read some articles on flying saucers and picked up a few books on the subject. I got hooked on the subject, more on a kind of fantasy level than anything else. I read about UFOs primarily for recreation and leisure, not necessarily because I believed that this was a real phenomenon and that we were being invaded by aliens. However, as I read, I began to realize that there may indeed be a 'signal' in the 'noise.' It took me awhile to get to that point, but I did get there. By the late 1960s, I was hooked and became very interested in the subject. It was then that I began to read seriously, and I subscribed to some UFO periodicals. By 1970, I was very much interested in the subject and was even writing my doctoral dissertation on UFOs. In 1975 my dissertation eventually became my first book, *The UFO Controversy in America*.

"My career, in essence, started with the UFO phenomenon. I didn't become involved with the abduction phenomenon until years later. Of course, we all had heard about the Barney and Betty Hill case. There were other cases here and there, but it was difficult to tell what had really happened. Most UFO researchers believe that the Barney and Betty Hill case could easily be psychological in its origin. And with just a few abduction cases at that time, there was no evident pattern. UFO researchers are primarily looking for patterns, with confirmation, verification, and all of that. This was certainly not the case with Barney and Betty Hill originally. It was interesting and exotic, but it wasn't a sighting. At the time, sightings were the focus of my attention. Sightings were what it was all about."

Gallup polls about UFO sightings have been consistent over the years. Since the 1950s, anywhere from 7 to 14 percent of the

American public acknowledges having seen a UFO. In a 1991 Roper poll of 5,947 people, 7 percent had seen a UFO, 5 percent dream of UFOs, 8 percent had puzzling scars, and *13 percent had missing time*. Missing time is, of course, a hallmark of an alien abduction experience. Information such as this must be scrutinized extremely carefully because missing time may also be a consequence of a psychological condition, for example, fugue states. Missing time may also be a result of memory loss from alcohol or drug abuse. Yet *the association of sightings with missing time* is significant and requires open-minded evaluation. If only a small percentage of the people reporting both sightings and missing time are being abducted, the numbers are quite large. For quite some time, though, sightings were "the only game in town." But that changed dramatically.

Dr. Jacobs noted that "Ultimately, though, what sightings amount to are people seeing the outside shells of objects. That's what you are looking at for the most part, except for a few occupant sightings. Most of the sightings we had by 1980—by which time we had hundreds of thousands of them from around the world—did not give us any insight into the purposes or motivation of the phenomenon, if it were extraterrestrial. It was an enormously difficult intellectual challenge to understand what the motivations and purposes could be, based on the sightings of the outside shells of objects, and the occasional sightings of occupants."

There were a number of early hypotheses concerning extraterrestrials, from the period of ufology dominated by and focused on sightings. The *Hostile Human Hypothesis* suggested that aliens were visiting Earth because of our violent behavior. The *Noninterference Hypothesis* implied that human society was so primitive that the highly evolved aliens would not interfere to avoid causing disruption and chaos. The 1950s "contactees" promoted what is called the *Space Brothers Hypothesis*, which viewed the aliens as our benevolent, wise teachers. However, there was little one could say about the motivations and purposes of the aliens based on sightings of UFOs. This changed with Dr. Jacobs's involvement with the abduction phenomenon.

"In 1982, I was introduced to Budd Hopkins by a mutual friend. Hopkins had already been doing abduction research and had published his book *Missing Time*. When I met Hopkins, he showed me what he was doing and introduced me to some of the abductees he was working with. I even went to an abductee get-together. I was extremely impressed with his work. It wasn't all what I had imagined. By 1983, I hadn't paid much attention to the abduction phenomena. I hadn't even read his book. I realized that, in order to investigate the UFO phenomenon and the abduction phenomenon, I would have to learn hypnosis. So I learned how to do hypnosis. Starting in 1986, I began to do my own regressions of abductees. So far, I have done over 700 sessions with over 120 individuals since 1986."

One enters murky waters when using hypnosis to gain access to repressed memories in the course of even completely human endeavors. How much more uncertain and tentative the information revealed must be when dealing with memories that may have been blocked by extraterrestrial intervention. It is useful to recall the recent controversy over the use of hypnosis to uncover memories of childhood sexual abuse—and the terrible damage done by false accusations of sexual abuse based on false memories. That being said, the valid, competent use of hypnotherapy has revealed patterns of alien-human activity that deserve thorough scrutiny.

Dr. Jacobs observed, "The question is, how do you separate the wheat from the chaff with hypnotically retrieved memories? This is a difficult situation. One of the problems UFO researchers have had has been that what people say in hypnosis is not always true. People *will* say things that are not true; they will confabulate. They can be wrong. There is such a phenomenon as false memory. There are also extremely rare cases where people will lie. Although a couple of these cases are well known, it is extremely rare that people will lie. But in 99.99 percent of all cases, people are not lying; they are just wrong. I went through a period where I was fumbling around, trying to understand what was happening, before I worked out a methodology that allowed me to be more confident in the data that I was getting. And to question more seriously and more closely what I was hearing."

A picture began to emerge for Dr. Jacobs through the information provided by the abductees. It was of an alien abduction program that seemed to be a breeding program of some kind. The focus on human sexuality, the removal of sperm and eggs, and the creation of fetal hybrids was reported over and over by the abductees. Could such extraordinary events actually be taking place? If one accepts the possibility that such a program could be underway, one is faced with the question of the seemingly enormous magnitude of the endeavor.

If the poll numbers are even remotely accurate, there could be hundreds of thousands, if not millions, of abductees in the U.S. alone. Could this be possible? We are in the infancy of our own human "breeding" programs. We are just beginning to experiment with cloning and, for over a decade, have been creating babies in test tubes. On February 1, 1998, the *New York Times* reported in its Sunday magazine that "about 10,000 human embryos are frozen in tanks of liquid nitrogen every year . . . *they're conceived in test tubes and stored for possible later implantation into a woman's uterus.* [my emphasis]" This is nearly identical to what the abductees report is happening. And their reports predate the ability of human beings to engage in these activities. If such things are occurring, the question as to why this is happening still remains.

Dr. Jacobs shared his thoughts on the subject. "I wrote *Secret Life* back in 1992, and now six years later *The Threat* has come out. About eighteen months ago, I was sitting in the cafe at one of the Borders bookstores here in the Philadelphia area. I was organizing my transcript notes about what the aliens said to abductees about what was going to happen to us in the future. I tape my hypnotic sessions and have them transcribed. I then put them in the computer into different slots, depending upon what they are talking about. For example, I have a large data base of examinations. And there are also large data bases of x, y, and z. One of them was a database of discussions of the future. I collected all those and put them out on the table. I began to look at them, read through them, and figure out what was going on.

"For the first time, in over thirty years of research, I began to get scared. I began to become frightened. I felt waves of fear coming over me. *I have never been frightened about this subject before.* Except for one fleeting moment years before, when I was discussing the scope of the UFO abduction phenomenon with Budd Hopkins. Budd was telling me about all these radio shows he had been on. He had been getting abduction calls from all around the country. We began to get an inkling of the enormous scope of the phenomenon. That scared me a bit at the time. I have to admit it made me very uneasy. But this time, while sitting in the bookstore, I was actually frightened as I began to read through the material. I realized that all of these people *were saying essentially the same thing, regardless of who they were.* None of these people were aware of what anyone else had said.

"I didn't like what they were saying. I began to realize that this whole alien program has a beginning, a middle, and an end. It was goal-directed. When the abductees talked about the future, they were all essentially saying one thing: *in the future, they— the aliens—will be here with us.* We weren't getting many other reports. If this were psychological, we would be getting a spectrum of future predictions, each one based on the person's own life and culture and society. Or maybe based on science fiction. Some people might say they are going to be here; some would say they will be on other planets. There might be a thousand different scenarios that people could come up with for the future. But actually, people were coming up with basically one. When they described what aliens said to them, there was basically one theme and one theme only: *in the future, they will be here with us.* Everyone is going to be happy. It's going to be wonderful. Everybody's going to love it. We will all know our places. We will all know our positions. It's going to be much better. I began to read over this material and I thought, 'I don't like this; I don't want this; this is no good.'

"Almost everything in the abduction phenomenon, from what I can tell, seems to point in that direction. Once again, if this were psychological, we would be all over the place. The reports would point in a thousand different directions. This is a reproductively oriented program, based on the hybrid phenomenon. There is a reason why the aliens are producing these hybrids. We are increasingly seeing abductions being *conducted by hybrids.* And we are seeing increased involvement of hybrids and people. All the evidence does seem to point in that direction. Quite frankly, I don't like it."

The abductees provide information when hypnotized. This information is then gathered, evaluated, and interpreted. The seemingly systematic nature of the abduction program, its global scope, and its apparent exploitation of the human beings involved could leave one suspicious of its purposes. Dr. Jacobs sees an abduction program that is a breeding program to develop hybrids, a kind of *Homo Alienus.* But is Dr. Jacobs's interpretation the only valid one, or the most likely one? There are other researchers in the field, whom Dr. Jacobs refers to in *The Threat* as the "Positives." These individuals—investigators such as Leo Sprinkle, Richard Boylan, Joseph Nyman, Dr. John Hunter Gray, and Dr. John Mack—view the aliens as our teachers. They believe that human beings are destroying our planet and the UFO and abduction phenomena are in some way involved with helping us learn to help ourselves. According to Dr. Jacobs, the Positives acknowledge the trauma inflicted on the human beings involved in alien abductions but consider it the price we must pay for growth.

As an open-minded investigator himself, Dr. Jacobs has written in his book that it is premature to say that the Positives are completely wrong. But he believes the facts support his viewpoint. "You have to remember that I do a tremendous amount of work on this in great detail. The average person comes to my house and leaves five hours later. The sessions themselves might be two or three hours long. I am very, very cautious; careful in my questioning; and very critical in what information I accept or do not accept. There are some things people say that I am suspicious about. Maybe it's not true. I wait for verification from others. Hopefully, two or three or four more people will verify it before I go forward. I have accounts of 'independent hybrid activity' and of 'personal project hybrid activity' and things like that."

In *The Threat*, Dr. Jacobs takes the alien abduction-breeding-hybrid program one step further. He believes that he has possibly uncovered—through the evidence presented by the abductees he has worked with—the ultimate goal of the alien abduction program. According to Dr. Jacobs, the abduction phenomena leads to the creation of hybrids who will be integrated into human society. *The abduction program is an integration program.*

"You have to remember that this is something that a number of abductees have been saying. It's not like one person here or there. When I first heard it, I couldn't believe it. I thought it was confabulation, that it was ridiculous. But in fact, I think it has become an important part of this phenomenon. And the first time that I have talked about it is in *The Threat*. It is the first time anybody has talked about it. I feel very confident in this data, as 'far out' as it may seem right now. I feel that other people will catch up with it eventually. I'm not sure that I perceive this alien program as hostile, though. I think that might be going too far."

Is this new idea so unthinkable? Dr. Jacobs recalled a story from the early days of the UFO phenomenon. "In 1946, at lunch one day, Enrico Fermi, the physicist, posited the famous 'Fermi paradox,' namely, if there are extraterrestrials—where are they? This was in 1946, before we all knew about the UFO phenomenon. His point was that because we don't see them, because they are not flying around, they don't exist. In 1975, a British scientist independently posited the idea that, if there were extraterrestrials out there, we would see an alien colonization program. They would be colonizing the galaxy or the solar system, just as mankind colonized the world. He said that's what we should be seeing with extraterrestrials. We are not seeing that so, therefore, there are no extraterrestrials here.

"However, I think, ultimately, that the UFO phenomenon and the abduction phenomenon answer the Fermi paradox. The question is, 'Where are the aliens?' and the answer is, 'They are here.' It also answers the problem of why they aren't they colonizing. I think the alien program is an *integration program*— which could be considered a colonization program. But it is certainly an integration program. Now, the aliens don't think that they are hostile at all. They don't think it's bad. They think everyone will be as pleased as punch and it will all be terrific. That's their point of view. From my point of view, I don't like it. I don't want them here. I would rather lead my own life and have my own society make its own mistakes. But *hostility* is different. Hostility means that they are angry; that they are coming here to do us harm. In their minds, they are not. I think the net effect might be harmful. There might not be any way to get around that, with the clash of a technologically superior society with a technologically inferior society. But I don't see them as necessarily hostile.

"When we take a look at this future society, one essential element that bothers me is that we are seeing a hierarchical society, with authority spreading downward from the top. That is something that all abductees indicate once you get into what the alien society is like. Once you put the accounts together, you see this kind of hierarchical society. The aliens themselves indicate that this is the kind of society we will be living in. A society in which they say everyone will know his position. Everyone will know his job. Everyone will know his place. I don't see the free society with the tremendous amount of individuality we now have, at least in the United States. I see a very differently organized society, more along the patterns of the alien's own life and society. Frankly, I don't like that and I don't want that. I am perfectly content the way I am right now. Once again, everything I am talking about stems from the evidence I have seen. Although it is speculative, it is evidentiary-based speculation, as opposed to just brainstorming."

In one way, human history is a story of failed attempts by people to create a better society. There have been magnificent attempts throughout the ages by Hindus, Buddhists, Jews, Christians, Moslems, agnostics, atheists, Marxists, and Capitalists to create a world in which human beings can prosper. Can anyone say that these earnest attempts have indeed resulted in a world we can be proud of? So, here we stand, lost in confusion, at a cosmic crossroads. Our very survival depends on the choices we make as we cross that intersection. And now we hear of yet another attempt to transform human society—an alien effort.

Could it be that, back in 1947 or 1957 or 1967, the clandestine group, or MJ-12, or whomever was collecting and secreting away UFO information, also came upon the information that Dr. Jacobs is presenting? Mere knowledge of UFO reality would pale in comparison to the discovery of a programmatic effort to biologically re-engineer humanity. Surely Dr. Jacobs must have considered the role the government might have played in this area. He did, and his answer is unequivocal.

"MJ-12, in my opinion, is a hoax. I think that most UFO researchers believe that. There are very few who don't think it's a hoax. I think that there is absolutely no evidence whatsoever for the idea that the aliens are appearing here for us to see and to desensitize us. That may be true; it may be what is happening. But there is no evidence for it whatsoever. Right now it is a product of pure conjecture, speculation, and guesswork. It is not evidentiary based in any way.

"Also, I find no evidence whatsoever for the idea that any responsible leaders here in the United States are trying to desensitize us by allowing little pieces of information to come out. I find that idea is not supported by facts, not supported by the evidence. Nobody has been able to come forward with any kind of chain of evidence that something like that has been going on for the past half-century, through ten presidents and ten administrations. To me, it is once again pure speculation, pure guesswork.

"Now, you can't prove a negative. Maybe there are some people hidden in the shadows who are pulling the strings. But I don't think government is that clever or that powerful. I think that most people in government are pretty much the same as everyone else. I really don't see thousands of people working on this over the past half-century and nobody knowing about it. Maybe I am a lot more cynical—or more realistic—about what government is and how it actually operates."

Although a covert campaign by the government to re-educate the public concerning extraterrestrial life does not seem realistic to Dr. Jacobs, he does not feel the same way when it comes to the ability of the aliens to engage in a successful covert campaign of their own.

"The evidence for the alien program indicates that it is clandestine and they want to keep it that way until their purposes are accomplished. They have worked very hard to maintain an enormously successful program of secrecy. The program has a beginning, a middle, and an end. One of the abductees I work with said that the program would be gradual, then accelerated and then sudden. It is hard to tell. But I think we are in the end stage now. I think you could say that this is an accelerated stage. But I think we are long past the beginning stages, and well past the middle stages. This is a program that has been going on now, we think, for about one hundred years. We can date it with certainty to the 1930s. With relative certainty, we can date it back to the 1920s. And with a degree of certainty, to the 1890s. That is the era of the 'Great American Air Ship Wave.' People argue over the Air Ship Wave. Many consider it mythical or folklorical. But actually, I think the alien program comes from about the same time. Before that time, however, we can't date it at all.

"It is important for people to know that there is a serious body of scholars who are actively researching this subject, and have been for years. There are maybe two hundred fifty to three hundred fifty scientists and academics interested in the UFO phenomenon around the country. It is not a nonsense subject. It is not going to go away. We are in for the long haul with this. This subject has nothing to do with events in our society. Because Nixon resigned that didn't mean people rushed out and saw UFOs. Or because Bill Clinton had troubles as President doesn't mean that people felt they were being abducted. This is a phenomenon that has a life of its own, unrelated to anything going on in our society, and it is certainly worthy of attention. I think that we should begin to pay very serious attention to it.

"This is a grassroots phenomenon where you have thousands of people coming forward, trying to tell the authorities or the scientific community or whoever they can get a hold of that *something is happening here*. It is happening around the world and it is unique. There has never been anything quite like this. You have to understand that the abduction phenomenon is not just stories. People see other people being abducted. They may or may not be abducted themselves. People are missing from their normal environments when they are abducted. Police have been called and search parties have been sent out. There is all sorts of physical evidence in support of the anecdotal evidence. You just can't dismiss something that is so all-pervasive and so potentially important. If anyone is looking at this for the first time, it is important to know that this is not a nonsense subject.

"I don't think there is anything that we can do about the alien program. But there is one thing that gives me hope: the fact that the alien program is still clandestine, still secret. That means they still feel that secrecy protects them against their vulnerabilities. Therefore, that means there is still a way of interceding. However, it is still secret and we don't know who most abductees are. We don't really know what is happening, and I don't think we will be able to find out. I didn't want to come to these conclusions. But I have been pushed into them by the weight of the evidence."

If Dr. Jacobs is correct, the alien abduction–breeding–hybrid-integration program will transform all aspects of human society—religion, science, government, education, family life, sexual life, and our view of who we are and what our place is in the Universe. Can a program that is carried out in secrecy, and that causes such trauma, lead to transcendence? The jury is still out. And it is we who are on trial, awaiting the verdict.

## The Best Available Evidence

Don Berliner is the Chairman of the Fund for UFO Research, an aviation historian, and a science writer who has published over three hundred articles and twenty-four books on aviation history, sporting aviation, space, and UFOs. He is is the co-author of *Crash at Corona* and the primary author of "Unidentified Flying Objects Briefing Document—The Best Available Evidence," which was funded by Laurence S. Rockefeller. The copyright for the report is held by the UFO Research Coalition, which consists of the Center for UFO Studies, the Fund for UFO Research, and the Mutual UFO Network. Intended for limited distribution to a select group of influential individuals worldwide, the briefing document may soon be available to the public through a major publisher. It is a powerful presentation of truly impressive evidence for the reality of UFOs and extraterrestrial life.

Berliner was "present at the creation" of the modern UFO era. He has been in the field of ufology since 1952. From 1965 to 1968, he was a staff writer with NICAP, the National Investigations Committee on Aerial Phenomena. "My initial awareness of UFOs was a pretty shocking experience," he said. "It did not come from a sighting but from a magazine article I read in December 1949. The author was Donald Keyhoe, and it was awfully well done and very persuasive. It stopped me dead in my tracks. I wasn't totally convinced that he was right. The idea that UFOs were real, rather than imaginary or optical illusions, came over me gradually. As did the probability that some of them were alien. There were no sudden shifts in my thinking after that article. It was all very gradual.

"I see UFOs as 'nuts-and-bolts' physical objects, some of which appear to be from outside the Earth. To me, that's the simplest answer, the most obvious. Some of the other answers require real stretches. In many cases, they get into the realm of science fiction. That doesn't interest me in the least. Likewise, I have no interest in the paranormal. My hands are full with just physical craft."

As a young man, Berliner's views on UFOs were influenced by Donald Keyhoe, the author of *Flying Saucers Are Real*. In fact, Keyhoe's thinking on the subject shaped the views of a generation of Americans. He proposed and expounded on the Extraterrestrial Hypothesis in his popular books and magazine articles. But Berliner has been around long enough to know that the UFO story is far from over.

Regarding his views on UFOs as physical objects from beyond Earth, he continued, "Now, I am not going to say that this is the total answer. We are still seriously short of information and certainly of proof. But that's the way it looks to me at this point. And it seems to answer more of the questions than any other explanations I have run into. Not all of the questions, but more of them. It seems to be a useful answer.

"I don't do a whole lot of research. I am not a scientist. And I think the term 'research' in the UFO field is horribly misused. People who do little more than clip out newspaper articles and put them in a scrapbook call themselves researchers. But I became more than casually involved—on more than just a hobby level—in 1952, as part of the Air Force's Ground Observer Corps. Before we had a huge network of long-range radar equipment across Canada, to detect approaching Soviet bombers, there was a stopgap effort called the Ground Observer Corps involving many tens of thousands of volunteers who stood on rooftops and out in fields with binoculars and telescopes looking for Soviet bombers.

"Of course, they never saw any Soviet bombers. But in 1952, there was a big UFO sighting wave in July. And this Ground Observer Corps produced a great many UFO sightings. They were out there looking, and that's what they saw.

"These reports went into any one of about fifteen centers around the country where they were plotted on a big map. There was a center across the street from where I was working part-time. I was waiting to get back into college after getting out of the service. I started monitoring the sightings. I managed to fast-talk my way into the center, though I had no authority of any sort. I collected the reports and passed the information along to the few UFO groups that existed in those days. That lasted for a couple of years.

"I did not become involved again until about 1959, when I got my first newspaper job in northeastern Ohio. My city editor found out about my interest in UFOs and funneled all the weird phone calls to me. When I moved to Washington, D.C., in 1962, I started working as a volunteer for what was then the world's largest private UFO agency, NICAP, the National Investigations Committee on Aerial Phenomena. I have been involved ever since."

Anyone who has seen a true UFO perceives it as an unforgettable experience. It is so out of the ordinary realm, and completely without explanation. Generally, people remember their UFO sightings vividly. In the early days of ufology, the main focus was on daylight and nighttime sightings. The more exotic aspects of the UFO phenomenon had not yet been discovered. However, today the focus of UFO research is shifting.

According to Berliner, "Sightings have become a less important part of the subject. Most sightings are at night, and people can't produce much information. In the good old days, we were getting a spectacular daylight sighting at the rate of one a week for years. Today, a spectacular daylight sighting would stop the presses. It used to be that we were not only getting spectacular reports, but they were coming from airline pilots, military pilots, that sort of thing. We just don't get much of that today. The major areas of interest today are abductions and government, with abductions being very exciting and government being not very exciting.

"We organized the Fund for UFO Research in August 1979. The Fund is composed of a fifteen-member board, most of whom are Ph.D. scientists. We do not have members or produce a newsletter or anything like that. We organized the Fund for the primary purpose of raising money to support scientific research and educational projects. We receive proposals, and the executive committee looks them over first. If the executive committee okays a proposal, it then goes to the full board. If the full board accepts it—and if we have money in the bank—we fund it. But we have also gotten into other things. We deal a lot with the press. We work with some people on Capitol Hill. We have an offshoot group that works with Washington, D.C., area abductees. We get information funneled to us and occasionally get involved in the investigation of a sighting. But our main activity is the funding of proposals."

The Fund for UFO Research and Berliner were involved in the publication of a UFO report considered controversial by many in ufology. At the end of his book, *Alien Agenda*, author Jim Marrs writes of this report—"Unidentified Flying Objects: The Best Available Evidence"—with a suspicious tone. Why, he wonders, was it prepared for only 1,000 members of the ruling international elite? Marrs believes that those in power may have a vested interest in keeping certain UFO-related information secret, especially knowledge of new, nonpolluting energy technologies which they could not control and exploit. Is he correct? Was the report prepared for a ruling elite interested in knowing about UFOs in order to maintain the status quo and its own position of privilege? One emerges with a different view after listening to the primary author discuss the purpose of the report.

"There is a group that is not terribly well-known called the UFO Research Coalition, composed of the Fund for UFO Research, the Mutual UFO Network, and the Center for UFO Studies. We work together on certain funding projects. A woman who was handling UFO matters for Laurance Rockefeller came to us with the idea for the report. As an individual, I made a proposal. It was accepted, and I ended up writing most of the report. The copyright is owned by the UFO Research Coalition. Rockefeller's only involvement was financial. He didn't even approve the outline or anything like that.

"This report has gotten to a great many influential people, but we haven't seen much of a result yet. However, we knew all along that the chances of anything happening quickly were very slim. The idea was to educate people in high places; people who, if they were sufficiently impressed, could make a difference. We have learned to be patient. It still may pay off big, but it hasn't yet. The report had a different purpose from any other UFO book. It was aimed at people who knew very little about the subject— and who had very little time. They are busy people. You don't give them a big book to read. We have gotten some very good comments from some impressive names but as yet, it has not turned the world on its ear. We stirred up a bit of a hornet's nest with the report because UFO enthusiasts could not get copies. But that's not what is was for. However, we are hoping to publish it commercially now."

Over the decades, Berliner has noted a shift in thinking about UFOs. While he feels that his thoughts on the situation today are close to the views he held in 1949 after reading Keyhoe's article, he has witnessed a change in public attitudes over the past fifty years.

"When the American UFO era began in 1947, nobody was thinking 'aliens.' They were thinking maybe the Soviets, maybe us. It wasn't until Keyhoe's first article—which was quickly followed by others and then by his book—that people started to think of aliens, not on a comic book level, and not on a hypothetical level, but on a very real level. I suspect that most people, when they hear 'UFO,' now think 'aliens.'

"This occurs whether they have done much thinking on the subject or not. People have gotten used to this over a period of years. The first time it hits you, it's a shock. The majority of the American people don't remember a time when UFOs were not connected to aliens. They are too young. People have become accustomed to this. If somebody with sufficient stature were to announce that some UFOs were alien spacecraft, most people wouldn't be that surprised. They have had an opportunity to get used to it. It's a big shift, but very gradual."

How has this shift come about? Is it a natural process related to the development of human society? Is this shift in thinking related to the discoveries that human beings are now making as we explore space? Or has there been, as the Mindshift Hypothesis postulates, an organized campaign to help people gradually accept the reality of extraterrestrial life?

Berliner said, "I don't see any real evidence of such a campaign to re-educate the public. And I don't see any need for it. The first person to push the alien explanation was Keyhoe. He fought tooth and nail with the government. I knew Keyhoe and worked for him for years. I can't imagine him being involved in that way. And anything that the government would tell him that was pro-alien, he would check out. He was a hell of a reporter. And he had tremendous contacts. The idea of such gradual, organized acclimation has been around for thirty years or more. I don't see any evidence of it. If somebody can point to some examples, fine. But I'm not impressed by the government's ability to be subtle, frankly. I've seen little evidence of it in any area.

"Of course, Keyhoe suspected things, such as leaking information. We all do. But he didn't talk much about that that I can recall. Sure, things happen that can be interpreted as an organized campaign, but that doesn't mean you are on the right track. For example, the recent behavior of the Air Force in relation to Roswell—the blindingly stupid reports the Air Force has been putting out—can be interpreted as a very clever way of getting people to take the whole thing seriously. But—at the expense of making the Air Force look foolish? I don't think so. The Air Force crash-test dummies explanation was the looniest thing it has ever done on the subject. The guy who wrote that report got canned. Not publicly. But we knew him and followed his brief career in the field with increasing amazement. I can't imagine that report as being part of a subtle effort to plant other ideas because it made the Air Force look bad. No government agency is going to purposely do that.

"*But I think it would make sense to prepare the public.* It would be a lot easier to do so now than many years ago. For one thing, our success in space has had a big impact on public attitudes and the public's ability to deal with shocking news. Remember the first Moon landing? Everybody stayed up all night to watch it! Did everybody stay up all night to watch the second Moon landing? No! It was old hat already. I think people are very flexible, have great resourcefulness and can handle things, as long as there is not a clear danger involved. And there is very little evidence of that in UFOs."

To all who accept the reality of UFOs as extraterrestrial spacecraft, no question is more alluring than *Why are they here?* And no answers are more elusive. Are they intergalactic imperialists, colonizing our planet? Are they wise, Ascended Masters, come to Earth to be our teachers? Are we the students who now are ready, and who have found our teachers? Or are they living, intelligent beings—rooted in Nature, as are we—who are exploring this portion of the cosmos, as we have heretofore explored only our beautiful blue planet?

Berliner said, "I have given a lot of thought to who the aliens might be and why they might be here. But it all comes back to the same place—we don't know beans about their psychological makeup. How do you interpret the behavior of intelligent individuals when you don't know how they think? You can't. If we did some of the same things the aliens have been observed doing, we could interpret that behavior. When our astronauts first get to another planet where there is life, they are going to do certain things. For instance, if there is just plant life, they'll collect plants and bring them back. They would be motivated by curiosity. But that doesn't mean that intelligent nonhumans would react the same way.

"It's terribly tempting to feel that if that's the way we think, then everybody must think that way. But it's pretty hard to justify. I think that anybody who talks seriously about the motivations of the aliens has an awfully big ego. I think it is extreme anthropomorphism. It cannot be justified.

"I wish we knew more about them. Interesting things have come out from abductees, but we don't know whether we should believe them. We are dealing with people whose memories have been tampered with. But we don't know to what extent they have been tampered with. Maybe it's a simple matter; maybe it's complex. For instance, it is entirely possible that all of the stories told by abductees about the interest of the aliens in genetics and reproduction could be a cover story for their real purpose. The aliens could plant all of that. We don't have any techniques yet to determine which is which. If we are getting the straight stories from abductees, that's one thing. Then we could learn something. But we don't know if we are getting the straight story.

"One of the really frustrating aspects of this whole business is that we are almost one-hundred-percent ignorant. It doesn't feel good. But it's the truth. Until you face that, you are going to come up with conclusions that can't be supported.

"When thinking about the UFO phenomenon—*keep an open mind.* Not only when you first get into the subject, but the entire time. Be very, very careful about drawing conclusions. *Make your own decisions.* Just because you see something on television doesn't mean it's true. Don't latch onto somebody else's opinions because they sound good. Simply weigh every new piece of information carefully. Even if you come to conclusions, don't be afraid to re-evaluate those conclusions because of new information."

Berliner's sage counsel regarding the UFO phenomenon has much broader applications, it seems. How different our world would be, and what pleasure and joy we would feel, if we journeyed through life with open minds, thinking for ourselves, able to change as new experiences shed new light on the majesty and mystery of our being.

## Contact with Non-Human Intelligence

Michael Lindemann is a journalist, the president of the 2020 Group, and the author of *UFOs and the Alien Presence: Six Viewpoints.* He is also the editor of the twice-monthly electronic newsletter *CNI News*. "CNI" stands for "Contact with Non-human Intelligence." This subscription newsletter addresses UFO phenomena, claims of alien-human contact, space exploration, and related issues, such as the cultural and political implications of contact with extraterrestrial intelligent life.

Lindemann first took a serious look at the UFO phenomenon and the possibility of alien presence in and around Earth in August 1989. At first, he thought a cursory look into the subject would indicate that there was nothing to it. However, he soon realized that he could not dismiss the possibility that aliens were visiting our planet and possibly even interacting with human beings. This was deeply disturbing to him. As a result, in December 1989, he started the Visitors Information Project as a focus of his future-studies organization, the 2020 Group. In March 1990, he published his preliminary findings as *UFOs and the Alien Presence: Time for the Truth*.

According to Lindemann, the evidence compiled by dozens of researchers in the United States, Europe, Latin America, and other parts of the world demonstrates that the alien presence on Earth is real, substantial, and purposeful. In his opinion, there are a number of types of entities operating on our planet. He also believes that secret, elite groups within the U.S. government, and other governments worldwide, are aware of the alien presence. However, for reasons known only to them, these groups have engaged in a massive coverup of this unprecedented development in human history.

Lindemann does not doubt there is a secret group or groups acting behind the scenes to coverup accurate information that would prove the reality of UFOs and alien presence on Earth. However, his views on a possible re-education campaign concerning extraterrestrial life, conducted by such a secret group, are not black and white. They come in shades of grey.

"Is there or is there not an actual educational campaign?" he began. "I have to say, I don't know. If I could answer affirmatively yes—and prove it—it would be a career-maker. But I can't. What I can say is that, if in fact the research perspective on the UFO phenomenon—that a fraction of these phenomena do connote the activity of another intelligence on our planet—is true, then it would surely be known to be true by somebody in the various governments. And those people would be *obliged* to create the means for absorbing this unusual reality. In other words, *there should be such an educational campaign* if any of these phenomena actually connote another intelligence in our air space. If I were given the task of designing the campaign, I would do exactly what I see happening.

"Basically, what I see happening is what I term a 'cacophony of conflicting images,' which creates several different impressions in people. First of all, a general impression that it is now culturally and politically correct to assume that these phenomena are real. But this impression is over and against a contrary message that these phenomena are not proven to be real, that they probably mean nothing, and it is probably silly to think about them."

Michael Lindemann characterizes himself as a futurist first; a social analyst, second; and a journalist, third. He employs all the knowledge, skills, and techniques of these disciplines in his evaluation of the UFO and alien phenomena. His multidisciplinary approach makes his point of view especially valuable. Although he does not know if there is a campaign to re-educate the public concerning extraterrestrial life, Lindemann makes a good case for one that would simultaneously reveal and conceal the truth.

"This dissonance is extremely important for creating the proper level of *psychic numbing* to get people in a frame of mind where, if something dramatic were to be announced, people would say, 'Oh, we already knew that!' I think it is absolutely, critically important that a majority of the public be brought to a state where, in effect, their visceral reaction would be, 'We knew that. This is not news. This doesn't change anything.' The big challenge has always been, hypothetically, that the knowledge of the reality of alien presence will shatter reality and cause the collapse of civilization as we know it. It is quite possible that this was never true. But it certainly is a lot less true today than it was fifty years ago."

Lindemann believes that the public could withstand an announcement that alien life is a reality. He even thinks it likely that the public could absorb a dramatic UFO event that was beyond the government's power to control. It is his view that we are now prepared to acknowledge UFO reality without wholesale collapse. These scenarios, either a government announcement or a dramatic UFO event, are big-picture scenarios involving whole societies and cultures. What does Lindemann think would occur on the personal level?

"I ask myself how I would feel if a grey alien came through my wall right in front of me. We hear a lot of stories exactly like that. They are very in-your-face, deeply personal, and shocking. Even if nothing happens—I stay put, am not hurt, and the alien disappears in a blaze of light and fog—how would I feel? To be honest, although I have been at this subject pretty much full-time for ten years, I still cannot guarantee to myself or anyone else that I would not simply wet my pants. It would be extremely alarming. It's a visceral level of reaction that even highly intelligent, grounded people might experience. Any government that presumes to be in charge of a vast culture like ours cannot afford to leave to chance the possibility that there could be this kind of reaction. *There has to be some kind of program going on. It is necessary.*"

The question arises as to how such an educational campaign would be carried out. The use of the mass media immediately comes to mind. Lindemann feels that "If you look at the things going on in the media, it can safely be said that a lot of it is simply the media doing what the media does, particularly the entertainment industry. It is always looking for a new hook, a new thrill, a new fad. But there are also stories that have been circulated saying that certain movies have had high-level consultants. For example, both *Close Encounters* and *E.T.* have been referenced in this regard. Such stories have been told many, many times. But we can't prove them. We have no 'smoking gun.' But, all in all, I think our culture has been moved along, whether purely by accident, or by extremely orchestrated design, or by something in between. It has been moved along very rapidly, and I believe this is accelerating. I see it moving quite rapidly in the last three to five years."

According to Lindemann, there are new factors that make the UFO phenomenon a great deal more palatable to the public than it used to be a few years ago. This shift is occurring because of developments in a field that has long been known for its hostility to and contempt for the UFO phenomenon—*mainstream science.*

"In the past few months, I have given a lecture many times. The talk is called 'Closing the Gap Between Space Science and Ufology.' My argument is simply this: although the space science camp and the ufology camp do not speak to one another, except derisively, in point of fact, they are increasingly saying the same thing. Most especially, the language that is occurring today in space science is highly supportive of the central claims of ufology. However, no space scientist is going to jump up and say so. But simply look at what they are saying:

"One, there is life out there. Our job is to find it. That is the virtually unanimous opinion in space science today. Life is now expected to be found even in our own solar system.

"Two, intelligent life is out there, and we will hear from it any minute now. This is another thing you hear constantly in space science now.

"Three, *the light barrier is just a technical problem. It is by no means an absolute speed limit.* It doesn't break the laws of God or Einstein or anybody else to surmise that there is a technology available to somebody somewhere—*a technology that we might even discover ourselves*—that can go faster than the speed of light. Once you have made that argument, once you have grounded that in physics and equations, you have just about eliminated all of the rational arguments against the possibility that somebody could come here from someplace else.

"I believe this has already occurred in space. The argument is now underway in space science. We haven't yet achieved faster-than-light travel. But we have already achieved in science what you would call a *mindshift* into an intellectual regime where breaking the limit of the speed of light is no longer disallowed. *It is simply a technical challenge.*"

Nearly four hundred years after Kepler formulated his laws of planetary motion, our space scientists use those laws to guide our exploratory space vessels through, and even beyond, our solar system. Is the futurist in Lindemann soaring too high, Icarus-like, with speculation that Einstein has been surpassed in his own century? Perhaps not. Consider these words, written by Wilhelm Reich in 1954, ". . . *there is theoretically no limit to speed in cosmic space, except technically.*"

Between 1940 and 1944, Reich wrote a manuscript in German that contained advanced mathematical equations that relate to the problem of space travel. It was not until November 1953 that the UFO phenomenon attracted Reich's attention. He saw a connection between the visual descriptions of UFOs and of reports of their performance capacities and the *mass-free energy equations* he had worked out over a decade before. Reich theorized that the mass-free energy that fills all of space could be the power source for spaceships. The reader will be introduced to this aspect of Reich's work in more detail in Chapter 11. But could it be that space science is now coming around to a scientific viewpoint Reich proposed fifty to sixty years ago? It would not be the first time such an event occurred in the history of science. It is a significant but little known fact that Newton's work was not accepted until approximately seventy-five years after his death.

Lindemann discussed an exciting new NASA research project related to these issues. "There is a NASA project called the 'Breakthrough Propulsion Physics Project.' Though it is not a big-budget item, it is a major project. They held their first symposium in September 1997. It was by invitation only. Major physicists, engineers, and space scientists all gathered to ponder three basic questions: One, can we break the speed of light? Two, can we do it without propellant? Three, can we access unlimited energy from space?

"The energy from space being referred to is known as zero-point energy. It is now an accepted reality in physics. Zero-point energy is the means by which it might be possible to create a practical, long-range space vehicle. The limiting factor in the current regime of space travel is propellant. Propellant limits both speed and range. It takes more propellant than you can carry to achieve the velocity you need to get where you want to go. Everyone realizes this. In fact, that argument is trotted out as a final, definitive argument against faster-than-light travel. But this ignores the possibility that *you could take your propellant or your motive energy right out of the fabric of space as you go— and never carry any propellant at all on board.*

"That is a big breakthrough. And that breakthrough was one of the hottest topics at the Breakthrough Propulsion Physics seminar held in September 1997."

The fact that mechanistic scientists are seriously studying the technical problems concerned with using the energy of space itself as a motor force for spaceships that could travel at faster-than-light speeds is truly a mindshift of major proportions. Is this development in any way related to the possibility that the technology recovered from crashed alien spaceships may have been reverse-engineered? Do ET ships use zero-point energy technology on their spaceships?

Lindemann said, "I have no idea how ET spaceships function. I have no idea, personally, if ET spaceships even exist. Do I think that there is a reasonable argument to be made for the back-engineering of ET technology? In a word, yes. I have interviewed Philip Corso and found him to be an impressive man. His pedigree is easily checked. He's been where he said he's been. He's done the things he's said he's done. I think the man is telling the truth as he believes it to be. Why would a man at his advanced age of eighty-two become a pathological liar at the end of his life?

"There is another factor to look at regarding reverse-engineering: the American Computer factor and American businessman Jack Shulman. To me, Shulman is a mystery man, more so even than Corso. He came out of the blue. He claims to have a long pedigree, but it hasn't been checked out as much as Corso's. But he seems to be putting a great deal at risk: a thriving company, his reputation, etc. Why is he putting all this at risk to tell an utterly cockamamie story about alien spacecraft and transistors, etc. It's very odd. I have talked to the man directly. I can't figure it out, but I think it bears watching. If any appreciable fraction of what Shulman is saying turns out to be true, it will be a big, big story. Shulman and Corso are sort of parts of the same story. If we establish finally that indeed space craft did crash on Earth by proving that the transistor could not have been created unless it had been taken from an alien ship, and not by proving that all these UFO cases we have are real, that would be a wonderful irony for me. And I'm halfway expecting it will happen."

There are many stories about reverse-engineering of alien technology. And the experts differ and disagree on major points. What technologies may have been reverse-engineered? Lindemann can't say for sure. "There are claims that we humans are flying exceedingly exotic spacecraft or aircraft that are way outside the regime of known engineering and physics. There are claims that we already have flying saucers and anti-gravity machines that allow us to go back and forth to the Moon or even Mars on a regular basis. Claims of this nature turned up in Jim Marrs's recent book, *Alien Agenda*, to give one example. And that's a fairly respectable book. To be quite frank, I think such claims are an absolute crock. I don't believe we can do all that stuff. I see no credible evidence at all. I could be wrong, of course. I know that our aircraft technology is well ahead of what is publicly visible. That's a given. But I do not believe that we have made the rather enormous jump to antigravity in a practical way."

The UFO phenomenon presents an evolving enigma. In a sense, the more one learns, the less one knows. Certainly far more *information* is available than there was fifty years ago. In 1947, the very idea of the existence of intelligent life was challenging. The notion that such life was visiting Earth was threatening to many. Today, there is discussion of fantastic possibilities that would have been unthinkable just a short time ago—alien abduction of human beings; medical procedures performed on people aboard space craft; an alien genetic engineering program creating human-alien hybrids using human sperm and ova. The familiar quotation from the Grateful Dead song "Trucking" is apt here: "What a long, strange trip it's been."

Michael Lindemann's involvement with the UFO phenomenon has taken him far afield from where he was in 1989. "Many individuals in ufology have undergone a shift from a 'nuts and bolts' to a more 'spiritual' or interdimensional point of view. I think it is very common. In fact, one could even say that ufology as a whole has followed that curve more than any other curve. My curve is different. I have always maintained a respectful distance from mainstream ufology. I have managed to have a good relationship with everybody in the field. I show up at a good number of the conferences, usually as a speaker. But I have always preferred to characterize myself first of all as a futurist; second, as a social analyst; and third, as a journalist.

"It was as a social analyst that I first took a look at UFOs in the late 1980s. Before that time, I had never thought about them even once, except as pure entertainment fiction. In 1989, I was leaving a seven-year stint as the director of a peace organization in Santa Barbara, California. During my years there, and for several years before that, I had pretty much been a full-time peace activist. I spent a great deal of time looking at U.S.-Soviet relations and the arms race. Among other things, I became very familiar with what I called *the fingerprint or the signature of secrecy.* What does secrecy look like? Does it leave a trail? Does it leave a cloud of smoke that indicates there's a secret over the hill? Well, it does. You can develop a nose for secrecy, and I felt that I had developed a nose for it."

According to Lindemann, with the end of the Cold War and the collapse of the Soviet Union, there was a dramatic shift in geopolitics that seemed to require a change in U.S. policy. It no longer seemed necessary for the United States to maintain maximum nuclear vigilance. Yet, it did. And the U.S. continued to develop extremely sophisticated weaponry and other military technology as if nothing had changed. Lindemann asked himself, "What's wrong with this picture? Why doesn't it add up? I asked all kinds of people and a novel answer came back to me—look at UFOs. I was completely surprised, but it came from a source I deemed reliable. I thought I would look at UFOs for a week or two, blow them off, and get back to work. Well, ten years later. . . ."

Lindemann came to UFOs looking for a deeply covert rationale for U.S. behavior that did not seem coincident with the visible changes in the geopolitical situation. As it turns out, he did not find a link between UFOs and the puzzling continued American military buildup. He speculated that the sheer momentum of the military-industrial complex may have kept the machine going. At that time, the U.S. was spending nearly $300 billion annually on the military, an amount of money that many people did not want to give up. But his explorations brought Lindemann to investigate UFOs. What did he discover?

"The first thing I noticed was that *they really are secret.* When that became obvious to me, I was shocked because I had been paying attention and I missed this completely. That was my hook. I thought, 'Boy, these guys are really good and really serious because I didn't even know what they were doing.' From there, of course, it became a morass and very frustrating. But I have stuck with it. In the meantime, I have gone through many stages, starting with a rather clunky analysis such as, 'The U.S. knows everything. They are keeping it all secret! A bunch of bad guys and thugs are running this operation!' A lot of reasonably smart people without a lot of experience jump right into a bunch of simplistic conclusions. Well, you have to graduate from that school as fast as you can. You have to begin to entertain subtler possibilities: One, most of the government doesn't know anything about UFOs. Two, most of the scientists who deny the existence of UFOs are not liars or stupid. They are genuinely unable to fathom how these things could be true. Three, whoever the aliens are, they are a lot weirder and sneakier than we thought.

Four, maybe the aliens don't come from anywhere we would call another planet or another star. Maybe they come from much more exotic places than that."

Earlier, Lindemann said his curve was different than the trend he noticed in ufology as a whole; that is, the development from a nuts-and-bolts approach to a spiritual approach to the phenomenon. Yet, doesn't his last point indicate that he, too, is moving in that same direction? He continued, "Of course, I went through all of those stages. But I came back around to believing—and, in this sense, I am counterintuitive to the general trend in ufology—that there is absolutely nothing at all wrong with the plain old, garden-variety Extraterrestrial Hypothesis. To me, today, it is the best theory of who the aliens are—*physical beings.* They are much like ourselves, driving spacecraft which are basically technological extensions of what we already know how to do, coming from places not unlike our own planet, from a star not unlike ours, at a distance reasonably far away. But not so far away that we couldn't get there ourselves. I think they come from an area I tend to call our Cosmic Neighborhood, which I have pegged at approximately twenty parsecs, or sixty-five light years."

Lindemann suggested that many space scientists and astrophysicists are discussing this matter at present. And they feel that a radius of twenty parsecs could reasonably be considered to be our neighborhood, in cosmic terms of course. According to Lindemann, "Inside that space there are at least dozens, and probably hundreds, of habitable star systems. We already know of a fair number of planets inside that space. We haven't even begun to find them yet. If you couple these findings with the technology of faster-than-light travel, all of a sudden, the Extraterrestrial Hypothesis makes sense. It's not something impossible.

"We come back to Arthur C. Clarke's dictum that any sufficiently advanced technology looks like magic. We see aliens doing things that look like magic and, because of the *paucity of our imaginations* and because of *our unwillingness to allow science to be the final arbiter* in a phenomenon that we would prefer to invest with some kind of 'sacred quality,' we are ignoring the possibility that we may be looking at beings very much like ourselves, using a technology just a tad more advanced than ours.

"We are right at the edge of a genuine paradigm shift in the sense that, for the first time, we will place ourselves not within a national or even a planetary environment, but within a *cosmos*, as part of a *cosmic community*. This jump in our perspective is, historically, a very rare occurrence. It opens up vistas of opportunity that we don't even know how to think about. But, on balance, I believe it opens *exceedingly positive* opportunities for us. Although fear is understandable, it is not appropriate. We will find, in the long run, that this is a perfectly timed, natural evolution of our place in the Universe, which is necessarily coincident with the evolution of our awareness and our consciousness."

At night, far away from large cities, by a vast ocean, on the shore of a deep mountain lake, or in the silence of the desert, the shimmering stars seem to beckon. In every corner of the earth, human beings have long told, and tell still, stories of our homes in the sky, stories of the living beings from those points of light above who have come to our blue planet. Many of us may look up at the night sky with longing. Some may dream of using the fruits of science—as did Edgar Mitchell—to set sail for the stars. Others may look up at the heavens and, recalling the wisdom of the ancients, seek to attain a deeper, sacred connectedness with Nature. Whether the path to understanding is through science or spirit, the stars call to us all. What is our answer?

PROJECT MINDSHIFT

Photo courtesy Photofest.

*What lies beyond* Contact?

# CHAPTER SIX

# Mindshift in the Movies
Mindshift in the Movies
Mindshift in the Movies
Mindshift in the Movies
Mindshift in the Movies

O ften, we fear what we most long for in life. We dream of freedom, yet fear freedom. We desire to be independent, yet exchange autonomy for security. We shackle ourselves to organizations and individuals whose goals and interests may even be antithetical to our own. This conflict within human beings frequently comes to the surface in the various forms of therapy people engage in as a means of self-exploration. This dichotomy is also clearly evident in popular movies. As we sit in the dark, watching lights flicker on the screen, the movies we love both reflect and reinforce our views of the world and ourselves.

The movies and space travel have been linked since 1902 when Georges Melies released the first science fiction film, *Le Voyage Dans La Lune*, or *A Trip to the Moon*. Melies was influenced heavily by the novel *The First Men in the Moon*, by H. G. Wells, and by *From Earth to the Moon*, by Jules Verne. The comic film was all of fourteen minutes long. But it was a commercial success, and the movie industry has been in love with outer space ever since.

In 1936, Buster Crabbe and Jean Rogers starred in a thirteen-part serial, *Flash Gordon in the 25th Century*. Although space travel was generally considered an impossibility at the time, audiences loved the daring couple who traveled in Dr. Zarkov's spaceship from adventure to adventure. The serial was followed by two films, *Flash Gordon's Trip to Mars* (1938) and *Flash Gordon Conquers the Universe* (1940). The success of Flash Gordon inspired a spinoff series, *Buck Rogers*. It was all good fun, and no one seriously considered the possibility of space travel.

In the five years after the release of *Flash Gordon Conquers the Universe*, the world witnessed the second global conflict in a generation, the Holocaust, the development of rockets, and the explosion of atom bombs over Hiroshima and Nagasaki. In the words of the poet Yeats, all was "changed, changed, utterly." To paraphrase Albert Einstein, the atom bomb changed everything except people.

A few years after Kenneth Arnold's UFO sighting and the Roswell incident ushered in the modern era of UFOs, the first Hollywood movie to use the term flying saucer in its title was released, *The Flying Saucer*. The title of this 1950 film was misleading. The plot of the movie revolved around, not an alien spaceship, but an advanced human airplane. However, the Hollywood studios recognized a hot topic when they saw one. By 1950 and 1951, people were increasingly reporting sightings of strange disc-shaped objects in the skies. More and more, the public was becoming convinced that these objects were extraterrestrial space craft.

In 1950, *Destination Moon*, produced by George Pal, directed by Irving Pichel, and adapted from Robert Heinlein's book *Rocketship Galileo*, was released. In an interesting sidelight, Heinlein had consulted with the German rocket scientist Hermann Oberth—who believed UFOs to be extraterrestrial spacecraft—about his story. Both *Destination Moon* and a poor-quality spinoff, *Rocketship X-M*, were financial successes. In 1951, George Pal made *When Worlds Collide*. This film, which won an Oscar for its special effects, told the story of the struggle of a few Earth pilgrims to build a rocket ship that would take them from Earth before it collided with another planet and was destroyed.

*The Thing from Another World* (1951), the only film ever made about an intelligent extraterrestrial vegetable, opens with an American military team finding a crashed saucer buried in the Arctic ice. There are hints of Roswell here. Inside, frozen, is an

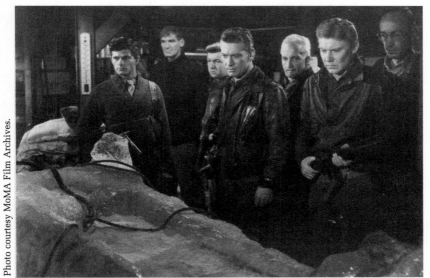

*Do elements of the Roswell story appear in the sci-fi classic,* The Thing from Another World?

eight-foot-tall alien. The alien comes to life and is discovered to be a being with no emotions and no sexuality. It fights desperately to survive, in the face of both the fierce climate and its human adversaries. The scientist at the base tries to save the creature—which has killed some of the soldiers because it feeds on their blood—but the military men eventually prevail and destroy it.

Other films involving alien beings, flying saucers, and space travel released in the early 1950s included *Superman and the Mole Men* (1951), *This Island Earth* (1951), *Invaders from Mars* (1953), and *It Came from Outer Space* (1953). The Superman film is interesting as a curiosity. *This Island Earth* has an intriguing theme—Earth is better off keeping a low profile; space is a dangerous place, and humanity may not be quite ready for what's out there. *It Came from Outer Space*, based on a script by Ray Bradbury, stands out because it is one of the first 3-D films. The story involves an astronomer who discovers that the meteor he witnessed crash in the southwestern American desert is, in fact, an alien spaceship. The extraterrestrials are no threat to humanity. They are completely focused on repairing their ship and heading home. Looking back in time, there are echoes of the

Roswell incident in the plot. However, there is something much more interesting about the other film in the group.

In *Invaders from Mars*, a young boy witnesses a flying saucer burrow into the ground in his backyard. His father investigates the next day and returns a changed man, with a black mark on the back of his neck. Soon, all the adults in town—including his mother—have the mark on their necks. It appears he has no one to turn to. However, the boy is able to convince an astronomer that his story is true. In the end, the aliens are vanquished. What is fascinating about the movie is that *the American version is different from the European version*. In Europe, audiences watched the film and saw the boy's story told as if it were *real*. In America, the audiences were shown the boy's story *as if it were all a dream*. The original screenwriter was so outraged by the tacked-on American ending that he had his name removed from the credits. Did someone intervene to make the American version match government policy?

In 1956, two classic science fiction films were released, *Forbidden Planet* and *Invasion of the Body Snatchers*. In *Forbidden Planet*, which is loosely based on Shakespeare's play *The Tempest*, repressed human emotions—particularly sexual desire, jealousy, and anger—wreak havoc in a world once inhabited by a superior race of beings who have vanished but have left behind a technology that turns the energy of pure thought into action. The repressed inner emotions of the humans—particularly the father's conflicted sexual feelings about his daughter—take the form of a murderous monster and nearly kill everyone. At the last moment, the truth is discovered. It is also learned that the creators of the technology had been destroyed by it themselves.

*Invasion of the Body Snatchers* is truly about "alien-ation." After drifting through space for millennia, alien seed pods find a suitable home on Earth, in the small California town of Santa Mara. They establish themselves on Earth covertly and, gradually, alien entities take over the bodies and the lives of the humans in town. The humans who are taken over become completely unemotional. Overnight, they have become strangers to their spouses, families, and friends. They have lost the capacity to love and be loved. Some people sense that a profound change is occurring, but they are too frightened to admit it or comprehend it. The alien message is delivered in the film by the town psychiatrist, who is

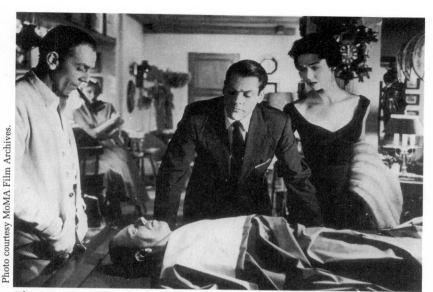

The Invasion of the Body Snatchers *forshadowed UFO themes to come.*

now one of them. He tells the hero of the film—the town's family physician—that the aliens are creating "an untroubled world" in which "everyone is the same." As the film ends, some humans are beginning to realize the danger they face from the alien invaders.

*I Married a Monster from Outer Space* was released in 1958. In this surprisingly moving story, a newlywed discovers that her husband is acting very strangely. He doesn't remember how to do the most ordinary things. It is revealed in the film that her husband is not the man she was engaged to; he is, in fact, an alien. He is a member of a doomed race that left their planet, came to Earth, and is now mating with Earth women in an attempt to preserve their species. With the help of the townspeople, the female protagonist manages to save her husband and defeat the aliens.

Some of these movies foreshadowed themes that emerged later in real-life investigations of the UFO phenomenon. In *The Thing,* the alien is perceived as having no emotions and no sexuality. Researchers who work with abductees have reported that aliens appear to lack both emotions and sexual organs. The story line of *It Came from Outer Space* has elements that are similar to aspects of the Roswell and other crashed-disc stories. The fact that the European and American versions of *Invaders from Mars* differ indicates that someone made a decision to present the UFO sighting as *unreal* in the U.S. version of the film. Who? And why? The change in the film reflects the debunking policy of the U.S. government at the time. *I Married a Monster from Outer Space* deals with a controversial theme, especially for its time—sexual relations between Earth women and alien males. This aspect of the 1958 film mirrors a theme developed in the 1998 book by David Jacobs, *The Threat*.

These movies were produced by the Hollywood studios to make money. They capitalized on the public's interest in flying saucers. To many people, such movies about aliens are no more than a subgenre of ordinary monster or horror films. To science fiction fans, these movies may seem to be attempts to explore and explain the Universe in a fictional genre they enjoy. Others view these films about space aliens as continuations of a folkloric tradition which tells tales of little people, fairies, wizards, and denizens of the netherworld. In all instances, the films are considered to be only entertainment.

Movies are money-making enterprises, but that is not all they are. They embody subtle and not-so-subtle cultural and ideological messages about family, about citizenship and patriotism, about the belief system of the people of the nation. Movies can be powerful tools to communicate information that indirectly promotes a hidden agenda. In this sense, movies are of great interest to people in power and were of even greater interest in the days before the rise of television.

Movies and politics have always mixed well because Hollywood and Washington are both about money and power. Hollywood moguls and Beltway power brokers use actors and politicians as performers to gain and maintain their fortunes. The power of, and effective use of, the movies was known to both Franklin D. Roosevelt and Adolf Hitler. FDR persuaded America's beloved, Oscar-winning director Frank Capra to take charge of America's film propaganda effort, and Capra produced the series, *Why We Fight*. Hitler persuaded one of the great filmmakers, Leni Riefenstahl, to use her talents to advance the Nazi cause. She produced the frighteningly powerful propaganda films, *Olympiad* and *Triumph of the Will*. The techniques she pioneered in those movies are taught at film school and are now often used successfully in a contemporary form of propaganda—

"hip" television commercials.

Presidents have frequently surrounded themselves with movie stars when running for reelection. Endorsements by beloved actors always enhanced a candidate's appeal. Hollywood is also a wonderful place to raise campaign contributions. John F. Kennedy enjoyed the company of Hollywood's "Rat Pack"—Frank Sinatra, Dean Martin, Sammy Davis, Jr., and Peter Lawford, the President's brother-in-law. President Kennedy also allowed the makers of the Hollywood film, *Seven Days in May*, unparalleled access to the White House and other government locations usually kept restricted. The movie, released in 1964, was later seen to contain plot elements that were eerily similar to circumstances surrounding the president's 1963 assassination. And President Lyndon Johnson had his own Hollywood advisor, the well-connected Jack Valenti.

The Hollywood-Washington Connection reached its peak in the triumph of style over substance with Ronald Reagan. President Kennedy made history when he became the first Roman Catholic elected President of the United States. Ronald Reagan was the first Hollywood corporate spokesman—or logo—elected President. In times of crisis, the vice-president has stepped in as *acting* President, as Nixon did for Eisenhower. With Reagan, America had its first *actor* President. In the 1990s, President Clinton has also been closely connected with Hollywood's movers and shakers, and turns to them for media and financial support. For example, filmmaker Steven Spielberg is a frequent guest at White House dinners honoring world leaders.

Why is the Hollywood-Washington Connection important in relation to the UFO phenomenon? Almost every president since the modern UFO era began has had to deal with UFOs in one way or another. It is alleged that MJ-12 began under President Truman and continued under President Eisenhower. Many stories—or rumors—suggest that President Kennedy was aware of UFO reality and wanted to be open with the American public about it, but could not. President Nixon is reported to have seen a UFO. President Ford, when he was a Michigan congressman, called for a Congressional investigation into UFOs. President Carter is on the record as having seen a UFO. And President Reagan frequently mentioned UFOs publicly on official occasions, such as in comments to Soviet leader Mikhail Gorbachev and at an address to the United Nations.

It is an axiom that politicians use the media, just as the media use politicians. Presidents, congressmen, senators, governors, mayors, and even city councilmembers all plant stories in the press. They all leak sensitive information that they want the public to know but that they cannot be associated with openly. For example, President Kennedy—a master at managing the news media—once complained to *Washington Post* editor Ben Bradlee about a story in his paper that made the Kennedy Administration look bad. Kennedy demanded that his friend Bradlee tell him the name of the person responsible for the leak. The seasoned editor had to tell the young President that the source of the embarrassing news leak was Kennedy himself.

Given the fact of the close relationship between Hollywood and Washington, it is highly probable that those in charge of the secret UFO information would try to use whatever means they could, not only to conceal UFO information from the public, but also to get information out to the public. The movies and aliens were made for each other. The keepers of the UFO secrets would have had no trouble planting information, suggesting story lines, or acting as unofficial consultants. All of this activity could be done almost unnoticed at Washington cocktail parties or Hollywood pool parties. No large "conspiracy" would be required to use the movie industry in this way, just as no large "conspiracy" would be required to keep the UFO information secret.

One film, released in 1951, not only appealed to and entertained a public fascinated with flying saucers, but also helped shape public attitudes about the reality of UFOs—*The Day the Earth Stood Still*. It stands out head and shoulders above the other flying saucer movies of the day. Nearly fifty years after its release, it remains one of the best flying saucer movies ever made. In fact, this film advances a thesis that is quite similar to the thoughts of ufologist Stanton Friedman.

Over the years, UFO debunkers have asked frequently, "Why don't the aliens just come to Washington and land on the White House lawn?" Well, in this film, they do. A huge spaceship comes to Washington, D.C. and lands, in broad daylight, on a baseball field in a public park near the White House and the Capitol. When the very human spaceman, Klaatu, finally emerges from his ship along with a huge robot named Gort, he finds himself ringed by the army, facing hundreds of weapons in

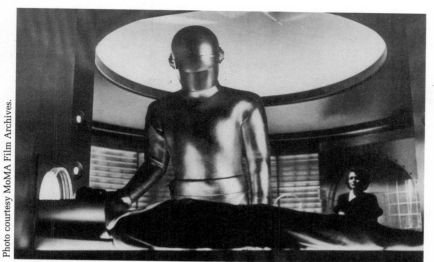

*Advanced energy technology set* The Day the Earth Stood Still *apart. Here Life Energy technology is used to bring Klaatu back to life.*

the hands of terrified soldiers. He reaches into his uniform and is shot when he takes out an unidentified object. He is surrounded by military men and tells them that the item was a gift for the President, with which he could have studied life on the other planets. The alien is rushed to a military hospital.

In the film, Klaatu escapes from the hospital, assuming the identity of a Mr. Carpenter. He befriends a young single mother and her son. Klaatu moves among ordinary Earthlings, seeking to learn what people are like. In one moving scene, the young boy, Bobby, takes the spaceman to Arlington National Cemetery to visit his father's grave. Looking out over the endless rows of white tombstones, the spaceman asks the boy in astonishment, "Did all of these people die in wars?"

Eventually, Klaatu is hunted down by the military and killed. However, he is retrieved by the robot Gort, who uses an advanced form of energy medicine to bring the dead Klaatu back to life. At the climax of *The Day the Earth Stood Still*, Klaatu delivers a message to a meeting of scientists from all over the world. Oddly, this film has the reputation of promoting a naive "Space Brothers" view of the aliens. Klaatu's message to Earth does not support that view of the film.

Here are the words that the spaceman speaks to all the peoples of Earth, whatever their station in life may be:

> *I am leaving now. You'll forgive me if I speak bluntly. The universe grows smaller every day and the threat of aggression by one group anywhere can no longer be tolerated. There must be security for all or no one is secure. This does not mean giving up freedom, except the freedom to act irresponsibly. Your ancestors knew this when they made laws to govern themselves and hired policemen to enforce them. We of the other planets have long accepted this principle. We have an organization for the mutual protection of all planets and for the complete elimination of aggression. The test of any such higher authority is of course the police force that supports it. For our policemen, we have created a race of robots. Their function is to patrol the planets in spaceships like this one and pre-serve peace. In matters of aggression, we have given them complete power over us. This power cannot be revoked. At the first sign of violence, they act auto-matically against the aggressor. The penalty for pro-voking their action is too terrible to risk.*
>
> *The result is that we live in peace without arms and armies, secure in the knowledge that we are free from aggression and war. Free to pursue more prof-itable enterprises. We do not pretend to have achieved perfection, but we do have a system and it works. I came here to give you these facts. It's no con-cern of ours how you run your own planet, but if you threaten to extend your violence, this Earth of yours will be reduced to a burned-out cinder. Your choice is simple, join us and live in peace or pursue your present course and face obliteration. We shall be waiting for your answer. The decision rests with you.*

After delivering his message, Klaatu takes off in his ship and returns to the cosmos.

This film has a message unlike the other science fiction movies of its day. In addition, it portrays a technology that far surpasses the Flash Gordon–Buck Rogers approach and special effects of other 1950s science fiction movies. Some of the elements of *The Day the Earth Stood Still* may cause some to wonder how those who made the film were able to present a technology that not only was unlike other films of the era, but also resembles technology that has actually been developed over the past half-century.

In the film, Klaatu returns to his ship to prepare for the event that gives the film its name. He is going to neutralize all the electricity on Earth for thirty minutes to capture the world's attention. He will bring human society to a standstill. When aboard the ship, he does not push buttons, pull levers, or turn knobs to activate the craft's equipment. Instead, *he uses the energy field of his body to turn on and work the energy technology of the spaceship.* This is a sophisticated concept, far beyond science fiction movies of the time—or the science of the day, for that matter. How did the filmmakers come by inspiration or information so far ahead of their time?

In another example, near the end of the movie, Klaatu is hunted down, shot, and killed by the military. Gort, his powerful robot, retrieves the dead body, returns it to the spaceship, and places the alien corpse in an advanced energy medicine machine. There are no levers or pulleys or buttons or other mechanical devices to be seen. Klaatu's body is suffused with radiation from head to toe, at ever-increasing levels. The energy machine is turned off; Klaatu is breathing. Slowly, he sits up and then stands. He has been returned to life.

As humanity enters the twenty-first century, medicine is standing at the threshold of the new discipline of functional energy medicine. It is to be expected that many things that would now be considered miraculous—such as the restoration of life after physical death—will become routine medical procedures. We already are able to perform similar "miracles" in mechanistic medicine. People who once died are now regularly resuscitated after a heart attack or revived after prolonged periods of hypothermia. In a similar fashion, it is to be expected that advances in physics— especially in the study and application of what is called *zero-point energy*—will yield equally astounding results.

Let us use the Mindshift Hypothesis here. If we assume that responsible leaders were aware of ET reality and presence, and if we assume that an MJ-12-type group existed and was in charge of such information, what would those in power do? If Corso and others are correct, they would keep secret information that was still volatile and threatened national security, while simultaneously trying to get other vital information out into the public's consciousness. We know that at that time, the U.S. and other governments used the powerful movie medium for propaganda purposes; to advance war ideology, for example, or to promote the use of atomic power. Would not the keepers of the UFO secrets use the film industry for their own purposes? The Hollywood-Washington Connection was already in place.

Is it possible that the makers of *The Day the Earth Stood Still* were given this information, either officially or unofficially, by people with knowledge of the technology of advanced alien civilizations? In any case, this film shaped public perception of intelligent extraterrestrial life in the 1950s and, through its availability on videotape, continues to exert its influence today. (It is interesting to note that, as this book was being written, *The Day the Earth Stood Still* was featured in the "Employee Picks" section of Channel Video on Manhattan's Upper West Side, indicating that a new generation holds the film in high regard nearly fifty years after its release.)

The 1960s were not years in which great science fiction movies were made about space aliens. In fact, such films were not doing as well at the box office as they had in the 1950s. Reality was both challenging and exuberant in the '60s, with the free-speech movement arising in Berkeley; the civil rights movement making great strides following Martin Luther King, Jr.; the peace movement growing each day as Americans turned against the Vietnam War; and the youth culture, with its message of peace and love, spreading worldwide, most effectively through the music of the Beatles.

But in 1968, the year before mankind reached the Moon, one of the greatest science fiction films of all time was released, *2001: A Space Odyssey*. Stanley Kubrick's masterpiece was not well-received by the critics. In fact, there was a good deal of hostility toward the film. But it benefited from great word-of-mouth advertising from the young generation coming of age in that era. The advertising campaign—which called the film "The Ultimate Trip"—clearly appealed to an emerging new mindset.

Kubrick was the director of *Paths of Glory* (1957) and the 1964 hit, *Dr. Strangelove: Or, How I Learned to Stop Worrying and Love the Bomb*. The term "Dr. Strangelove" had already become a part of popular culture. The same fate was in store for *2001*. The film, loosely based on Arthur C. Clarke's short story "The Sentinel," is a complex, intellectual film which dramatically deals with the intervention of an alien race in the development of mankind. A giant black Monolith—which in retrospect looks like it is made of the material used decades later in Stealth technology—appears in the film as the symbol of alien intervention that "jump starts" human evolution.

In *2001*, each alien intervention results in the development of a weapon by the humans, and in murder. The primitive Hominids, after a frightening encounter with the Monolith, learn to use bones as clubs with which to kill; after the humans on the Moon encounter the Monolith, they follow its signal toward Jupiter, and all but one—a space-age Ishmael—are killed by the advanced, thinking computer that controls the mission. Unlike Melville's Ishmael, however, he does not return to tell his story.

*2001: A Space Odyssey* certainly helped shape the mindset of a generation about space travel, the nature of mankind, the possibility of alien intervention in human evolution, and the mystery that awaits humanity as it explores the Universe. But again, there seem to be elements to the film that stand out as possibly being the result of more than pure creative imagination. *Paths of Glory* is an incisive study of war and its insanity. Its reputation has grown over the decades. *Dr. Strangelove* beautifully portrays the surreal rationality behind the military minds in both the U.S. and U.S.S.R., who proposed and perpetuated the nuclear policy of Mutually Assured Destruction, or MAD.

It is highly likely that there were individuals in government and the military who appreciated Kubrick's work despite, or even because of, its portrayal of political and military madness. In fact, the keepers of the UFO secrets may have been more likely than others to appreciate the difficult truths Kubrick was presenting. Those who knew the truth of ET presence in detail could not have avoided undergoing a personal "mindshift," whether they wanted to or not.

This is not unheard of in history. There is evidence that some at the highest levels of the Roman Catholic Inquisition were well aware, for example, of the truth of the work of Giordano Bruno and the accuracy of Galileo's scientific discoveries. However, to preserve the power of the Roman Church, and the societal structure of their day, they chose to oppose Bruno and Galileo in public. It was decided that the people were not ready to hear the new truths about the Universe to which they had access.

Information was released, through various unnamed government sources, in the years after the release of *2001*, suggesting that there was indeed alien intervention in human evolution in the distant past, approximately 30,000 years ago. The Air Force's own textbook, *Introductory Space Science*, is reported to state that ETs have been involved with humanity for at least 50,000 years. This book was in use at the Air Force Academy before *2001* was made, as was similar information. UFO researcher Linda Moulton Howe has reported that she has been shown documents that purport to detail ET involvement with humanity over similar periods of time, documents that appear to shed light on man's origins. Did the makers of *2001* have access to this information? An intriguing but, as yet, unanswerable question.

In 1969, a low-budget, youth-culture film took the United States and Hollywood by storm. *Easy Rider*, written by Peter Fonda, Dennis Hopper, and Terry Southern and directed by Hopper, told the tale of two alienated bikers off in search of America. The movie was awash in drugs, with the two main characters high throughout the film. *Easy Rider* was the movie that catapulted Jack Nicholson to stardom. Nicholson played an alcoholic attorney who joined up with the two counterculture bikers. In one classic sequence, the three men are sitting up late at night, smoking marijuana, in the deep stillness of the desert. At one point, Hopper sees a UFO. Nicholson proceeds to tell the others that aliens from space are living and working among the American people. He says that this is not only known to government officials at the highest levels of authority, but that the aliens and the government are working together.

This dialogue contains ideas that are "far out" even for the drug-suffused mindset of the counterculture as presented in *Easy Rider*. The two lead characters cannot accept what Nicholson tells them; Hopper is visibly upset and demands that Nicholson stop talking about such crazy things. The film was originally aimed at the so-called "Woodstock Generation" but it crossed over and was seen by a large portion of mainstream America.

In what way could the Hollywood-Washington Connection be at work here? In the real-life CIA project called Operation Paperclip, hundreds of Nazi psychiatrists were brought to the United States, where they continued their psychochemical brainwashing experiments. Psychedelic drugs—such as mescaline and LSD—were used in many of these experiments because they showed great potential for use in mind-control operations with the public, as well as with individuals. These drugs were also an essential feature of the world of *Easy Rider*.

According to Martin A. Lee and Bruce Shlain, authors of *Acid Dreams: The CIA, LSD and the Sixties*, one of those who came to America through Operation Paperclip was Dr. Hubertus Strughold. Despite recurring allegations that he participated in medical atrocities during World War II, some involving experiments with the psychedelic drug mescaline, Dr. Strughold took up an important position in the American space program. He was later acclaimed as the "father of space medicine" and was highly respected at NASA.

In the 1980s, in addition to Lee and Shlain, many other authors reported on the work of former Nazi scientists in the American space program and in secret CIA-sponsored psychochemical brainwashing experiments, some of which were done using psychedelic drugs given to unsuspecting members of the American public. Timothy Leary, the "Johnny Appleseed" of LSD, was even quoted as saying, "The LSD movement was started by the CIA. . . . It was no accident. It was all planned and scripted by the Central Intelligence."

One of the techniques the clandestine group allegedly employs in releasing UFO information is to present the truth in circumstances that undermine its believability. It is highly possible that such truthful information about alien life could have been suggested to the *Easy Rider* filmmakers. Certainly, during the '60s, Terry Southern, Fonda, Hopper, and Nicholson would have mixed at "hip" Hollywood happenings with all kinds of covert "spooks" and overt government officials. The success of the film—and the enthusiastic acceptance of its message by many young Americans—may not have been quite what the keepers of the secrets had in mind, however.

Steven Spielberg's two influential films—*Close Encounters of the Third Kind* (1977) and *E.T.—The Extraterrestrial* (1982)—

shaped the popular view of aliens in the 1980s and 1990s as much as *The Day the Earth Stood Still* did in the 1950s. More than any other filmmaker, Spielberg is associated with a "mind-shift" in public thinking about extraterrestrials. And there are more rumors about covert government involvement in his films than about any other filmmaker as well.

In *Close Encounters*, all of the themes of the modern era of ufology come into play—the reality of extraterrestrials and UFOs; sightings of ships and of entities; human abduction by aliens; a government cover-up campaign; and ultrasecret government contact with the ETs themselves. Spielberg presents not only the big-picture, conspiratorial elements, but also the human side of the UFO story. As exhilarating as contact with extraterrestrials is to the human beings involved, the experiences also leave them disoriented and troubled. The most significant aspect of the movie may be that, despite the government cover-up, the truth does come out because it is pursued, independent of any authority, by those who have been touched by the reality of the UFO phenomenon.

Spielberg brilliantly shows the UFO phenomenon to have been ongoing since at least World War II, and to be a global phenomenon. He portrays the kooky side of ufology humorously as well. And he shows that UFO reality touches men, women, and children from all walks of life. The director peers behind the veil of secrecy to see, not a malevolent government conspiracy, but an organized group of humans attempting to hide a reality that they do not yet understand.

When contact is made, it is a transformative cosmic event. A huge mother ship arrives at the climax of the movie. Abductees from the past fifty years are returned by the aliens, and present-day humans voluntarily join the aliens in the next stage of the human adventure. Interestingly, the alien-human rendezvous point is at Devil's Tower, Wyoming. The Native Americans living in that area have a myth about Devil's Tower that relates directly to the story of *Close Encounters*.

In the myth, a giant bear chased seven maidens, who sought refuge atop Devil's Tower. The bear tried desperately to get at the maidens, clawing away at the side of the tower in a futile attempt to climb to the top. In the myth, this is how the tower got its distinctive, clawed-surface look. The myth ends with the arrival of the beings who "seeded" humans on Earth. They return the maidens to the heavens, their origin. Today, in the myth and in the sky, we see the seven sisters shining above as the seven stars Western astronomy calls the Pleiades. Not surprisingly, there are many UFO books available about the involvement of extraterrestrials from the Pleiades with human beings. To the Native Americans who have long lived in that area, the site is a sacred place of worship.

In the film *E.T.—The Extraterrestrial*, a new generation of children and their parents were introduced to extraterrestrials. Unlike the movies of the 1950s, in which aliens were always a threat to humanity, E.T. does not frighten the children. He is kind, loving, and beloved by them. Neither is the alien omnipotent, as in most other movies. The alien, dubbed "E.T." by the youngsters, has been abandoned by his fellow beings, just as the young boy, the film's protagonist, has been abandoned by his own father. E.T. needs the young boy's help and protection. In the movie, the adults cannot see the reality of E.T. until it is too late. The children cannot protect the alien, and E.T. falls into the control of the military-scientific machine men. In the clutches of their lifeless technology, E.T. appears to die. However, the alien has deceived the humans and, with the help of the children, he escapes from the evil secret government forces and finds his way home.

*E.T.—The Extraterrestrial* has helped to create a mindshift among a younger generation around the world. Children saw that the extraterrestrials were not to be feared. They were living creatures, much like ourselves, despite differences in physical appearance. They were imperfect, living beings—not monsters determined to destroy, devour, or enslave humanity.

Before he made *Close Encounters of the Third Kind* in 1977, Steven Spielberg had directed the television movies, *Duel* (1971); *Something Evil* (1972); and *Savage* (1973). In 1974, he directed the film *The Sugarland Express* and, in 1975, he directed the blockbuster hit, *Jaws*. How did the director of such films come to make one of the most influential movies ever made about contact between humanity and extraterrestrial life? There is nothing in his past work that indicates he is a director interested in exploring any such cosmic themes or creating any such consciousness-raising stories.

Today, Steven Spielberg is a regular guest at White House dinners held by President Clinton. It is interesting to note that in 1995, President Clinton signed an executive order that may result in a great many now-classified UFO-related documents being released to the public. In addition, it has been widely reported that Laurance Rockefeller has been urging President Clinton to take steps to make available information about the early days of the modern UFO era, particularly from the Roswell period. There are also rumors that Spielberg may soon make another film with an extraterrestrial story line. It will be interesting to see, should such a film be made, if it contains UFO information that may have been passed in casual conversation at a White House fête or a Jackson Hole, Wyoming, brainstorming session.

In 1984 the excellent film *Starman*, directed by John Carpenter, was released. Jeff Bridges was nominated for an Academy Award for Best Actor for his portrayal of an alien who struggles to survive on Earth after his spaceship is shot down by the American military. The alien—first seen escaping his crashed ship as an energy being—makes his way into the home of a young widow living in a remote lakeside cabin in Wisconsin. Using advanced genetic engineering techniques, the alien takes DNA material from a lock of the late husband's hair, which the young woman has kept in a photo album, and creates a clone of the husband's body. At first, the alien kidnaps the woman, forcing her to take him to the meteor crater near Winslow, Arizona. As the movie progresses, the alien begins to feel what it is to be human. At one point he even remarks, "I think I am becoming a Planet Earth person." And the woman, Jenny Hayden, begins to love the alien, partly because he resembles her late husband, partly because of who he is.

In *Starman*, the love element grows into lovemaking. Love between an alien and a human woman are present in a restrained manner in *The Day the Earth Stood Still*. Sexual contact between aliens and Earth women was the main theme of *I Married a Monster from Outer Space*. Intergalactic love emerges in *Starman* in a beautiful scene where the spaceman and the Earth woman make love in a railway car.

In a moving scene, Jenny Hayden awakes after lovemaking to see the alien looking at her with deep affection. The dialogue between the two star-crossed lovers reveals many UFO-related themes that were not yet in the popular culture:

Jenny: *Isn't there any way you can stay?*
Starman: *No, I must go back. . . . But there is something I must tell you. I gave you a baby tonight.*
Jenny: *No, that's impossible. I can't have a child. The doctor said—*
Starman: *Believe what I tell you. A boy baby. He will be human. The baby of your husband. But also, he will be my baby. He will know everything I know and, when he grows to manhood, he will be a Teacher.*
Jenny: *Which is your star?*
Starman: *Why?*
Jenny: *I want to show him where his father came from.*

In the 1980s and 1990s, through the work of Budd Hopkins, Raymond Fowler, David Jacobs, John Mack, and others, the first stories began to appear of extraterrestrial interest in human sexuality; sexual intercourse between human beings and aliens; and the creation of alien-human hybrids through artificial techniques using human sperm and ova. The cold, indifferent, alien-hybrid program envisioned by Hopkins and Jacobs, based on the evidence offered by abductee reports, stands in contrast to the complex human-alien interaction described in Fowler's and Mack's work.

The alien in *Starman* is not an omnipotent being. He is merely another lifeform, one of the many that exist in the cosmos. As in *E.T.*, the alien needs the help of human beings to return home. If he does not meet with his rescue spaceship in time, he will die. His life is in the hands of a SETI researcher, affiliated with the National Security Agency, who does not seem, at first, to have the courage to help the Starman. The following, decisive scene, in which Jenny Hayden, Starman, and Mark Sherman of SETI interact, is quite revealing:

Sherman: *Is there anything I can do for you?*
Jenny: *You can let him go.*
Sherman: *I can't. Really, I can't. I'm sorry. Are you supposed to meet someone here? Is that it?*
Starman: *Yes.*
Jenny: *You don't understand. There isn't much time.*

Sherman: *Why here? Why the crater? Have people from your world been here before?*

Starman: *Before? Yes. We are interested in your species.*

Sherman: *You mean some kind of anthropologist? Is that what you're doing here?*

Starman: *You are a strange species—not like any other; and you would be surprised how many others there are—clever but savage. Shall I tell you what I find beautiful about you? You are at your very best when things are worst.*

The SETI/NSA investigator rises to the occasion and lets the alien go, that he may return home. His behavior is unlike the human approach to ETs in movies from the 1950s. In *Starman*, we again see ideas and concepts appearing in a mainstream Hollywood film that will later appear in the UFO literature and make its way slowly, over ten to fifteen years, into public consciousness.

In the summer of 1997, as NASA's Mars Pathfinder mission captured the imagination of the world, a breakthrough science fiction film was released, *Contact*, directed by Robert Zemeckis and starring Jodie Foster. Previously, Zemeckis had made the successful films *Back to the Future* (1985) and its two sequels, as well as the hugely popular *Forrest Gump*. In *Contact*, based on the novel of the same name by UFO debunker Carl Sagan, an intelligent message from an extraterrestrial civilization is received through the official U.S. government program SETI, or the Search for Extraterrestrial Intelligence, a program that Sagan enthusiastically supported. Although he dismissed the possibility of intelligent alien life visiting Earth today, the late astronomer Sagan told a 1966 gathering of scientists that ". . . the earth may have been visited by various galactic civilizations many times (possibly on the order of 10,000) during geological time." Sagan believed that alien visitation was possible in the distant past, but not today. Is that logical?

*Contact* deals dramatically with the consequences of ET reality for religion, government, education, and even interpersonal relationships. The film's release was certainly well-timed, coming in a year in which stars with planets were being found regularly; ancient fossils of microbial life were discovered on a Martian asteroid; water may have been discovered on Earth's moon; life seemed probable on at least one of the moons of Jupiter. Never before had life elsewhere seemed so abundant.

The intense longing for contact with life elsewhere was beautifully portrayed in the movie. And the conflicts among scientists, the military, politicians, and powerful behind-the-scenes forces, all struggling to control the event and its consequences, were handled extremely well. It remains to be seen if *Contact* is revealed, at a later date, to contain information about extraterrestrial life and our interaction with it that is now known only to a select few. If the past is anything to go by, it is highly likely that some material, seeded in an organized or offhand manner, has made its way into this movie.

The top-grossing American movie of 1997 was another UFO-related film, *Men in Black*. This comedy is a take-off on the more serious concept of Men in Black (MIBs), which has been around in ufology since the 1950s. These MIBs are unidentified agents who claim to be from unnamed security agencies. They frequently appear at the homes of people who have had UFO sightings or other interactions with aliens and threaten them with harm if they speak out about what they know. *Men in Black* found a huge summer audience but, as pure comedy, it did not advance human acceptance of the UFO phenomenon, except indirectly. However, its wide appeal indicates how deeply the subjects of UFOs and alien life have penetrated public consciousness.

As this book was being completed, a new book appeared that focused on a theme similar to the Mindshift Hypothesis. In Bruce Rux's *Hollywood vs. the Aliens: The Motion Picture Industry's Participation in UFO Disinformation*, the text, in some instances, seems to support the main assumption of this book, although the author has a different point of view. Rux writes that the main tenet of his book is that military intelligence and the CIA have exploited, and continue to exploit, the entertainment industry to get their messages across. But he focuses on disinformation and misinformation. However, Rux's viewpoint is in agreement with the Mindshift Hypothesis in some ways. He writes, in the Introduction to his book, that the purpose of these intelligence agents is one of "alternately confusing and enlightening the pub-

Photo courtesy Photofest.

*One of the aliens living among us in* Men in Black.

lic at large on UFO facts," a statement completely in line with the Mindshift Hypothesis.

Rux mentions a number of links concerning the Hollywood-Washington Connection that are relevant to the Mindhsift Hypothesis. For example, he reports that many UFO films—some containing UFO information not generally known at the time—were made by RKO Pictures. RKO was owned then by the billionaire Howard Hughes, who was connected to many secret CIA and military intelligence projects in his life. RKO itself was part of the empire of Henry Luce, founder of Time-Life, who was himself connected to the CIA and other intelligence agencies through CIA Director Allen Dulles. The publisher of *Life* magazine, appointed by Luce, previously worked as a CIA consultant on psychological warfare. According to Rux, Roger Corman of American International Pictures, which made many UFO movies, may have had intelligence connections.

William Joseph Bryan, Jr., a CIA agent whose specialty was hypnosis and mind control, was a top advisor on many of Corman's films and later on many television shows as well. Dr. Bryan's son, C.D.B. Bryan, is the author of a book, *Close Encounters of the Fourth Kind*, about the 1992 UFO Conference at MIT. In a thorough and fascinating book, the son deals with his father's CIA-UFO connection in only a few sentences. Rux also discusses other examples of the Hollywood-Washington Connection, such as that between Walt Disney and intelligence agencies, the "educational campaign on the UFO subject" that he feels occurred during the Carter Administration, and other fascinating topics.

The Mindshift Hypothesis provides an interesting tool that the reader can use to evaluate the UFO films of the past, present, and future. So many are available now on video, and it is relatively easy to get hold of them at the local video store or even by mail. Many movies were not discussed here—the *Star Wars* trilogy; the *Superman* movies; and the many *Star Trek* films, for example. The assumption that movies contain secret information is one that can be tested simply by watching the films with a critical eye, looking for the clues that are there.

Mass media such as the movies provide powerful means to reach a huge audience and shape and reinforce public opinion. It is highly unlikely that the clandestine group, the keepers of the UFO secrets, did not use movies to re-educate the American public concerning extraterrestrial life.

*Photo courtesy Photofest.*

*Television's most influential Mindshifters, Special Agents Mulder and Scully, shaped the pop Zeitgeist of the '90s.*

# CHAPTER SEVEN

# Mindshift on Television
# Mindshift on Television
# Mindshift on Television
# Mindshift on Television
# Mindshift on Television

A merica is a land of great myths. One of the greatest modern American myths is that the airwaves belong to the public and not to the media giants that control them. Is there even the slightest shred of evidence that the public truly owns the communications bands that ABC, CBS, NBC, Fox, CNN, TBS, and others use to accrue vast profits? Legally, of course, the airwaves are public property. But this legal fiction does not exist in the real world of media megabucks. The public airwaves basically now belong—all the rules and regulations notwithstanding—to a powerful few who are licensed to use them by the Federal government. The media barons need the government licenses in order to do business. Therefore, they are willing to do "business" with government officials.

In 1983, in his book *The Media Monopoly*, Ben H. Bagdikian reported that fifty media corporations controlled the production and distribution of most movies, television programs, cable television shows, radio shows, recordings, books, and magazines. By 1992, Bagdikian noted, in a revised edition of his book, that there were only twenty dominant media corporations. He estimated that, by the year 2000, a handful of powerful media conglomerates will dominate information in

America. And, as the popular saying has it, "Information is power."

Dennis W. Mazzocco, in *Networks of Power—Corporate TV's Threat to Democracy*, explored the media's links to government, the military, global banking, and financial institutions, and analyzed how the needs of these groups shape the messages sent out over the airwaves—and how these forces limit those messages. According to Mazzocco, concentrated media ownership leads to limitations on public knowledge. He sees an increasing, dangerous fusion between the television industry and government.

Viewers see instances of this quite frequently now. For example, David Gergen has been an advisor to Presidents Nixon, Carter, Reagan, Bush, and Clinton. He is also a regular commentator on PBS and a frequent guest on Sunday morning talking-head news shows. Pete Williams was the spokesman for the Pentagon under President Bush; he now works for NBC. George Stephanopoulos was an adviser to President Clinton; he now works for ABC.

This symbiotic relationship has had profound consequences for society throughout the history of the extraordinarily influential force that is television. The implications for the treatment of the UFO phenomenon on television are enormous. Powerful corporate and government institutions and individuals use the media to promote and reinforce their policies and agendas daily. In his book, *Hollywood vs. The Aliens*, Bruce Rux asserts that it would be inconceivable for the government not to use the medium of television for disinformation or educational purposes on the subject of UFOs and extraterrestrial life.

The modern UFO era and the age of television both date from approximately the same period. It is not surprising, then, that television has played a large role in both shaping and reflecting the development of public attitudes toward UFOs and extraterrestrial life.

The first television show about outer space was *Captain Video and His Video Rangers*, which was on the air from 1949 to 1955. In April 1950, *Buck Rogers* made its first appearance and lasted until January 1951. *Tom Corbett, Space Cadet* was an extremely popular show, with superior special effects for its time. It actually appeared on each of the three networks and was sometimes broadcast three nights a week. In June 1951, *Space Patrol* debuted on ABC. It met its demise in June 1952. The show was then produced and shown locally on the West Coast until 1953. *Flash Gordon* was also a popular show in the early 1950s, as was *Rock Jones, Space Ranger.*

Television was never embarrassed by its lack of originality, and this was seen in the great number of similar programs about space. Other entries from the era included *Red Brown of the Rocket*

*Captain Video in a classic pose.*

*Rangers*; *Johnny Jupiter*; *Jet Jackson, Flying Commando*; *Captain Z-ro*; and *Commander Cody*. However, the number one hit was *Superman*. From 1952 to 1957, the intelligent extraterrestrial from the Planet Krypton fought for "Truth, Justice and the American Way" in 104 episodes.

*Science Fiction Theater* was an intriguing program. Many of the episodes concerned UFOs. One show even centered around LSD. This drug was not well-known in the 1950s, but this was the era in which the CIA began testing LSD as part of its covert psychopharmaceutical brainwashing program.

These early television shows were primitive by contemporary standards and, in general, they were at a level of sophistication that was not much higher than a comic book. The situation changed dramatically on October 2, 1959, when the pilot episode of *The Twilight Zone* was aired. The series was an immediate success and remained on the air until 1964. It is seen regularly in reruns and on cable television to this day. The show was noted for its excellent writing, particularly the scripts by Rod Serling and Richard Matheson (who influenced Stephen King). *The Twilight Zone* had a deep impact on the public and played a major role in shaping attitudes toward UFOs and extraterrestrials.

One early episode, "The Invaders," told the story of a woman who was terrified by aliens who had come to her remote farmhouse. The "aliens" turn out to be American astronauts exploring another world. Another episode that featured astronauts was called "And When the Sky Was Opened." In this tale, astronauts return to Earth only to vanish, one by one, both physically and from all records and human memory. In "People Are the Same All Over," an astronaut is greeted on his arrival at another world by Nordic-type aliens, only to be put in a zoo by his human—all too human—hosts.

One of the classic alien episodes of the show is "To Serve Man." In this story, a race of tall, robed, humanoid aliens brings paradise to Earth. Only a few untrusting cryptographers in the military remain suspicious. They eventually break the code of an alien text. As the protagonist is boarding a flying saucer for a cruise to the alien planet, his assistant comes racing toward him, begging him to stop. Too late, she calls out, "To Serve Man—it's a cookbook!"

In another classic episode, "The Monsters Are Due on Maple Street," aliens bring chaos to a small town simply by turning the electricity on and off. The human beings panic, become suspicious, and turn on one another, destroying their own society. Martians, the occasional Venusian, and other aliens made frequent appearances on *The Twilight Zone* in such episodes as "Third from the Sun," "Probe 7—Over and Out," "Will the Real Martian Please Stand Up?" and "Mr. Dingle, the Strong," In an interesting turnabout, a Martian risks his own life to save a dying girl in "The Fugitive."

*The Twilight Zone* frequently manifested a cynical, jaundiced view of humanity as its own worst enemy. The show opened the minds of millions of regular viewers to the many mysteries that await them in the twilight zones of their own lives.

Another television show that had a big impact—although it lasted only two seasons—was *The Outer Limits*. Each episode began with an ominous voice intoning, "There is nothing wrong with your television set. Do not attempt to adjust the picture. We are controlling the transmission. . . ." The show premiered on September 16, 1963, with its pilot, "The Galaxy Being." In this episode, a man on Earth invents a 3D-TV that connects him to an alien in the spiral galaxy of Andromeda. The contact with the alien brings about tremendous fear in human beings; this fear of the unknown is the focus of the show.

In "Architects of Fear," an Earth man is surgically altered. He pretends to be an alien who has come to Earth to force the nations of the world to unite and work together in the face of alien reality. Things go awry from the start in this unusual look at humanity's view of itself and life from elsewhere. One episode, "Nightmare," revolves around an alien device that crashes on Earth and is then used as an excuse by the government to engage in its own covert, destructive activity. Insectoid aliens make an appearance in "The Zanti Misfits," and a merciful being from a world of light struggles against cruel, avaricious Earthlings in "The Ballero Shield."

Two *Outer Limits* episodes involve abduction scenarios. In one show, "Fun and Games," a human being is abducted by aliens, taken to their planet, and forced to fight for his survival against alien beings. In "A Feasibility Study," an entire city neighborhood is suddenly transported to an alien world by its own government. "The Chameleon" tells the story of an Earthling who allows himself to be biologically altered so he can infiltrate an alien ship and destroy it. However, the spy comes to understand and love the aliens and despise human beings. He flees Earth with the ETs.

"The Invisible Enemy" is set on Mars. An American crew is searching for an earlier Mars mission that has vanished without a trace. It slowly becomes clear that the danger to the humans lies beneath the Martian soil. *The Outer Limits* aired only one two-part episode, "The Inheritors." This plot line concerns men who are using alien technology to build a spaceship, which will take sick children to another world where they can heal and live happy, healthy lives.

What is striking about a number of these episodes from both *The Twilight Zone* and *The Outer Limits* is how they reflect official fears, such as mass panic upon learning of ET reality; how they correctly portray human fear of the unknown and the frequent destructive reaction to it; and how they anticipate things to come, such as the recovery of advanced alien technology and the abduction phenomenon. The exploration of human fear is a common theme in all forms of writing. But many of the specific plot elements in these shows—insectoid aliens, crashed alien technology, human abductions—were known only to the keepers of the UFO secrets at that time.

One wonders how the psychedelic drug LSD made its way into a show on *Science Fiction Theater* in the 1950s and how television plot lines contained information about aliens that would not enter public consciousness for decades. Are these simply examples of vivid imaginations at work, or did these ideas come through the Hollywood-Washington Connection?

The channels for the transmission of such information existed from the very beginnings of the television industry. The great corporate giants that would have been involved with covert government projects concerning the reverse-engineering of alien technology, for example, were all intimately connected to the television industry, as either owners of networks or corporate sponsors. The Mindshift Hypothesis could easily be at work. Again, no great conspiracy is needed. Ideas can be put out into the air and then float along, like the seed pods in the film *Invasion of the Body Snatchers*, until they find fertile ground in which to take root and grow.

It may be difficult to imagine today, but the most influential of all science fiction television shows, *Star Trek*, was not very successful in its day. It premiered on NBC on September 8, 1966, and remained on the air until 1969. The show had consistently low ratings during those three years. Today, *Star Trek* is broadcast over two hundred times a day in the United States alone. It has been dubbed into forty-eight languages, is a hit worldwide, and has spawned other television series and many feature films. There are about seventy million *Star Trek* books in print in over a dozen languages.

Many episodes in this series concern subjects, and highlight plot elements, that are now known to be essential features of the UFO phenomenon, but which were not common knowledge thirty years ago. For example, "The Cage" involves an abduction scenario in which aliens study human emotions. Another episode, "The Empath," involves what initially appears to be a cruel alien abduction involving the study of human emotions. Later, the alien activity turns out to be part of an attempt to preserve life.

In "Naked Time," contact with an unknown alien lifeform causes great emotional upheaval in the crew of the Enterprise. Spock, the emotionless Vulcan, is overcome by grief and weeps. Captain Kirk longs to let go of the shackles of authority and lead an ordinary life. In other crew members, repressed emotions sur-

Star Trek's *instantly-recognizable starship, the* USS Enterprise, *the vessel that has helped shift the minds of millions.*

face in the form of rage, lust, and intense fear. Today, abductees regularly report that the study of human emotions seems to be a key element in the alien abduction phenomenon as they experience and perceive it. But this information was not available when these episodes were written and aired.

The Starship Enterprise itself appears on Earth in the 1960s and is thought to be a UFO by the American military. In "Assignment Earth," the time-traveling crew of the Enterprise encounter a character, Gary 7, who was abducted from Earth as a child and raised by an alien race interested in Earth. Gary 7 is attempting to sabotage an American nuclear space program effort on the instructions of the aliens to protect humanity from the consequences of this deadly technology. In later years, in real life, many stories appeared concerning alien observation of civilian and military nuclear operations, and even possible alien alteration of thermonuclear bomb codes at military bases. In December 1980, in Britain, a UFO landed near an American nuclear base and contact was apparently made between the American military and the ETs. In another show, "Tomorrow Is Yesterday," the Enterprise once again finds itself back in time and is mistaken for a UFO. The plot of this episode was published in a revised form in a popular magazine for young students.

In *Cosmic Voyage*, Courtney Brown, Ph.D., writes about his remote viewing experiments as they relate to *Star Trek* in a chapter titled, "*Star Trek* and the ET-Assisted Transformation of Human

Culture." Dr. Brown is an associate professor of political science at Emory University, specializing in nonlinear mathematical modeling of social phenomena, environmental politics, and elections.

Dr. Brown claims to have made remarkable discoveries about ETs using the controversial technique of scientific remote viewing (SRV). This technique was developed by the military and used extensively by the CIA in a $20-million, multi-year program that the intelligence agency has acknowledged publicly. Brown was trained by a military expert in SRV. According to Brown, humans have been, and are now, involved with two alien civilizations, one from Mars, the other the so-called Greys. Brown also believes that there is an advanced "Federation" of civilizations to which Earth may one day belong.

*Star Trek: The Next Generation* often seemed to contain ideas and data that Brown was discovering through SRV contact with extraterrestrials. He began to wonder if the minds of the writers of the show were being manipulated by the ETs to help open up the human mind to the reality of intergalactic life. In this way, the mass medium of television could be used to help reshape public thinking. Brown's military teacher had long suspected that Hollywood shows were somehow part of an educational campaign.

Brown's SRV session led him to believe that extraterrestrials had put the idea for the show, or at least the idea of supporting the show, into the minds of Hollywood power brokers. Another SRV session brought him to the conclusion that specific *Next Generation* episodes were influenced by ETs. According to Brown, these ideas are put into people's minds, but they are not forced to accept them. There is no mind control or brainwashing involved. However, he firmly believes that ETs have influenced human culture through television via *Star Trek* and possibly other shows as well.

There is no way to prove or disprove Brown's assertions. It could even be that, with his military connections and unusual academic specialty, Brown himself is part of a covert human disinformation or education campaign that has nothing to do with ETs at all. Whatever the case may be, *Star Trek*, and its many film and television progeny present a positive message concerning humanity's place in the Universe and its interaction with the many lifeforms that inhabit the cosmos. Its overriding themes are a far cry from the fear and paranoia of the 1950s.

*The multicultural crew of Star Trek: The Next Generation reflects an optimistic outlook on future relations with alien races.*

In addition, *Star Trek* clearly focused on the need for human beings to develop, to evolve, before they could take their place in the cosmic community. The series conveyed neither a "Space Brothers" attitude nor a hostile enemy attitude toward intelligent life elsewhere. Humanity's fate was in its own hands and would be determined by its own actions.

*Star Trek: The Next Generation* picked up the banner and took viewers along with the Enterprise as it explored the Universe from 1987 through 1994. *Deep Space Nine* and *Voyager*, two more spin-offs, continue on the air today. But the show that has defined the pop zeitgeist of the 1990s—as *The Twilight Zone* did for the late 1950s and early 1960s—is the hit series, *The X-Files*.

This show is the highest-rated ever on the Fox network for viewers eighteen to forty-nine years old. It combines mystery, police drama, and science fiction in a series with a unique visual look and a unique philosophical outlook. In the 1960s, a popular motto was "Don't trust anyone over thirty." For fans of *The X-Files*, that slogan has been updated to "Trust No One."

Each week, Special Agent Fox Mulder (whose sister either has or has not been abducted by aliens) and his partner Special Agent Dana Scully (who herself either has or has not been

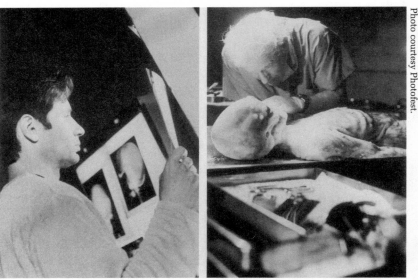

*Special Agent Mulder seems to have proof of ET reality in his hands.*

abducted by aliens) encounter life-threatening situations involving government conspiracies, alien conspiracies, and a wide range of paranormal phenomena. The name "Scully" harkens back to 1950 and Frank Scully, author of *Behind the Flying Saucers*—possibly the first victim of government disinformation.

The major difference between *The X-Files* and earlier shows that dealt with the UFO phenomenon is that it does so consciously and uses a wide range of readily available information. One does not need to look for clues in the plots. It is immediately obvious where certain elements of the show come from, such as the Roswell incident, alien abduction stories, or reports about Area 51, a secret government research center in Nevada.

The show went on the air in the 1993–1994 season and attracted about three million viewers, a small audience for television. However, by 1996 the audience had more than tripled to 10.7 million viewers. The number of viewers continues to grow and, by this writing, has reached about 13.7 million.

*The X-Files* forces the viewer to use his or her imagination. In the type of show the producers call "stand-alones," the plot is usually resolved by the end of the hour. However, there is another type of episode, called "the mythology shows," that deals with the ongoing interactions Mulder and Scully have with extraterrestrials and UFOs. After watching these episodes, viewers usually have more questions than answers.

In June 1998, the show makes the leap to the big screen with its first feature film, *Fight the Future*. There is also a cottage industry of *X-Files* novels and books about the show. To the viewers, government conspiracy and covert alien activity seem to be the big draw. However, many who work on the show find more enjoyment in the plain old "monster" episodes that are a regular part of the series. Creator Chris Carter was greatly influenced by a show from the early 1970s, *The Night Stalker*, and this is clearly evident in the "monster" episodes.

As with most influential television shows, *The X-Files* both exploits an existing public attitude, namely mistrust in government, and reinforces that attitude. In addition, the show exploits the ever-increasing public interest in extraterrestrial life and UFOs, and simultaneously reinforces it. The growing success of *The X-Files* may itself be evidence that the Mindshift Hypothesis of an educational campaign about UFOs is working. Rather than being a part of the re-education campaign, *The X-Files* may be a sign of its success.

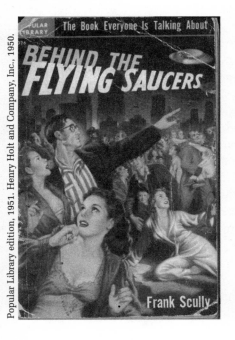

Popular Library edition, 1951. Henry Holt and Company, Inc., 1950.

Television dramas are not the only format in which a mindshift is clearly noticeable. Documentaries and even television commercials indicate that attitudes toward UFOs are changing. In the spring of 1995, the Walt Disney Company aired a syndicated television documentary entitled "Alien Encounters from Tomorrowland," ostensibly to promote a new ride, "Alien Encounters," at Disney World in Florida. In the opinion of *CNI News*, this documentary was devoted to basic information about UFOs, claims of alien contact, and substantive information about government cover-up of UFO and ET reality.

Oddly, the show aired at unusual hours with little or no advance publicity. According to *CNI News*, the film's narrator claims the viewer is seeing "photographs of an actual spacecraft from another world, piloted by alien intelligence, one sighting from tens of thousands made over the past fifty years on virtually every continent of the globe. Intelligent life from distant galaxies is now attempting to make open contact with the human race."

The documentary claims that the Earth is "experiencing a tsunami of sightings" and that they are not reported in the press because "For governments determined to maintain their authority, extraterrestrial contact is pure dynamite. . . ." The film asserts that, not only was there a crashed UFO at Roswell, it wasn't the only one. The film also treats the controversial clandestine group, MJ-12, as a reality. In one particularly astounding statement, the Disney documentary informs the viewer that "Statistics indicate a greater probability that you will experience extraterrestrial contact in the next five years than the chances that you will win a state lottery."

In its conclusion, Disney informs the viewer that, in "New Tomorrowland at Disney World," the company's scientists and engineers have brought to life a scenario to help the public adjust to an *inevitable alien encounter.*

On May 12, 1996, the A&E Network showed "Where Are All the UFOs?" This was a familiar debunking effort. A few nights earlier, The Learning Channel had aired a debunking documentary. PBS also entered the debunking arena with a *Nova* special on UFOs that dismissed the subject as nonsense.

However, the big advertising agencies have tested the ET waters and discovered that whether aliens exist or not, they help sell products on television. Commercials for washing machines, automobiles, and beer are now all over television, with cute aliens playing the lead roles. Kodak ran a commercial in which an ordinary American finds he has inadvertently caught a UFO on a role of snapshots he took while fishing. He has used a digital camera that captures the images on a disk. After viewing the UFO on his computer, he sends the picture out over the Internet. This alerts both the government and the ETs, neither of whom is pleased, both of whom come after the young man.

**Reverse engineered from UFOs.**

Drivers wanted

Aliens have been used in commercials for the *Star Tribune* in Minneapolis, Ding Dong snack cakes made by Hostess, Breathe Right Nasal Strips, and the newly redesigned Volkswagen Beetle. The Quaker Oats Company has an alien spokesbeing for its cereal, Quisp. Dan Geib, webmaster of the website "UFO Folklore," wrote an article titled, "Aliens in the '90s—Conditioning or Commercialism?" He wonders whether humans are being conditioned through television commercials, or if simple greed is the motive for the alien onslaught in advertising.

In July 1997, on the weekend of the fiftieth anniversary of the Roswell incident, the airwaves were filled with aliens. The Sci Fi Channel showed *Roswell, It Came from Outer Space*, and *The Blob*. The Discovery Channel aired a six-hour documentary marathon, *U.F.O.s Down to Earth*, and a news show, *Mars Live*, about the Pathfinder landing. At the end of July, Fox showed a rerun of *Alien Autopsy* and a documentary, *U.F.O.s: The Best Evidence Ever Caught on Tape*.

During the week of June 29 through July 5, 1997, in the New York City metropolitan area, the following UFO-related films were shown: *The Arrival; The Blob; The Cat from Outer Space; Destination Moon; Earth vs. the Flying Saucers; Independence Day; It Came from Outer Space* (1953); *It Came from Outer Space II* (1996); *Star Trek II, IV,* and *V;* and *The UFO Incident*. Similar programming was most likely scheduled in every television market in the country.

One of the most significant UFO documentaries is *UFOs: Past, Present and Future*, made in 1974 by Robert Emenegger and Sandler Institutional Films, Inc. Emenegger explains, in his book of the same name, that when he began work on the project, he was a *skeptic*. He thought all UFO sightings were the result of fantasy or wishful thinking. However, his mind was changed once he was exposed to the overwhelming evidence supporting UFO reality and ET presence.

The documentary is based on fact. It gives an accurate overview of the UFO phenomenon throughout history, from ancient times to the present. Interestingly, in terms of the Mindshift Hypothesis, the film was made with the cooperation of NASA and the Department of Defense. The UFO cases from the modern era were carefully chosen, thoroughly researched, documented, and then rechecked. Most of the cases were taken from Air Force files and had been investigated by intelligence agents from various government agencies.

*UFOs: Past, Present and Future* is unusual in its low-key tone; its use of official material and footage; the cooperation of the Defense Department and NASA; and the clear, direct manner in which it presents the evidence. And the evidence points in only one direction—UFOs are real and they represent, in some manner, an unknown intelligence that has entered our earthly existence for reasons as yet unknown.

More than at any previous time in its history, the television industry is under the thumb of global corporate interests, through advertising and direct ownership of networks. Disney owns ABC; General Electric owns NBC; and Westinghouse owns CBS. The latter two corporations are military-industrial giants who influence what Americans see and hear. For example, CBS and NBC rarely, if ever, air news stories critical of the nuclear industry that is so important to the parent companies. There are other such examples.

Many other important issues are subject to internal self-censorship within the media. NBC and CBS also frequently highlight launches and arrivals of the space shuttle. However, the networks ignore evidence of the great damage the shuttle causes to the environment. They also ignore the tremendous risks involved in the use of nuclear power in the American space program. Here again, the effects of corporate financial interests can be seen affecting news coverage.

Does the television industry cooperate with the government to conceal information? Look at the censorship the military imposed in 1991 during the Gulf War. The industry accepted it with little complaint. Does the television industry promote the government's political agenda at times? Again, look at the coverage of the 1991 Gulf War. ABC, CBS, NBC, and CNN turned a real war into a made-for-television movie. Each network had a martial theme song for its Gulf War show. Their art departments created special logos for the occasion. The television computer artists had more fun with high-tech video images of advanced weaponry than a grade-school child with a video game. High production value videos of the war were in the stores as soon as the hostilities had ended. Television made sure not to show the men, women, and children—especially the children—who were dying.

Would the television industry conceal UFO information if asked to do so by the government? Could it be used to plant stories about extraterrestrials that were false? Or used to launch trial balloons of accurate but volatile UFO information to guage public reaction? Would the industry collaborate with a clandestine group within government? Its history of interaction and involvement with government makes the answer clear.

Despite the fact that broadcast television still remains "a vast wasteland" in many ways, it has the capacity to do great good, to bring people together in wonderful ways. The television industry achieved something of this nature in its first global satellite transmission, which included a live performance of "All You Need Is Love" by the Beatles. Is it an impossible dream to believe that, one day, the medium of television could be used to broadcast the truth about UFOs and extraterrestrials to all of the peoples of Earth, simultaneously?

*Imagine. . . .*

*Was the press used to both reveal and conceal the truth about UFOs and ETs?*

The Curtis Publishing Company, 1966.

# CHAPTER EIGHT

# Mindshift in Print
## Mindshift in Print
### Mindshift in Print
Mindshift in Print
Mindshift in Print

# EXTRA
## TERRESTRIAL

O ver the past year, the *New York Times* has attracted the attention of its readers by running the above word in display type in unexpected places in the different sections of the paper as part of an advertising campaign for itself. Under the word "EXTRA terrestrial," the copy continues, "Extra coverage of the earth. And beyond."

Although the newspaper of record sees fit to print ads for its Science Times section exploiting the interest of its readers in extraterrestrials, the paper remains quite shameless in its determination to ridicule the subject whenever possible. In fact, the paper seems to create opportunities for itself to do so. For example, in early 1998 in "Photography View," in the Sunday Arts and Leisure section, there was a story headlined "Of Fairies, Free Spirits and Outright Frauds," about an exhibition on the famous Cottingley fairy photographs. The *Times* used the article about the early twentieth century British photographs to ridicule UFOs and extraterrestrial life as outright frauds.

The *Times* noted that "Otherworldly sightings today are generally depicted in decidedly down-to-earth ways: see Andrea Robbins and Max Becher's unassuming photographs of the area near Roswell, N.M.," and then continued, in cutesy, put-down prose, ". . . where space cadets believe the Government is hiding a U.F.O. that dropped by for a cup of coffee 50 years ago."

Aside from the fact that no one believes that the government is hiding a UFO in Roswell, the scientists, military officers, astronauts, cosmonauts, and intelligence agents who have information about crashed extraterrestrial spacecraft and who know that an alien craft, or possibly two, crashed in New Mexico in 1947, are not "space cadets." But the derisive tone, and use of derogatory terms such as "space cadets," characterize the editorial policy of the *New York Times* concerning intelligent extraterrestrial beings and their space craft. This is a policy that has held sway for nearly fifty years and shows no signs of changing.

In terms of the Mindshift Hypothesis, the *New York Times* plays a critical role. Its views reach and affect highly influential readers in crucial positions in government, media, religion, education, and other critical fields. Its news stories, syndicated through its own wire service, are reprinted all over the United States and the world. In addition to fulfilling its role as a cautious, responsible voice of reason in society, is it possible that this prestigious newspaper would cooperate with a covert government campaign concerning extraterrestrial life?

It is worth considering a story published in the *New York Times* on June 6, 1997, under the headline "Role of C.I.A. In Guatemala Told in Files of Publisher." According to this story, which the paper published about itself, "In June 1954, the publisher of the *New York Times* privately agreed with the Director of Central Intelligence to keep a foreign correspondent of the *Times* out of Guatemala, according to the publisher's personal files, just as the Central Intelligence Agency was secretly mounting a coup there."

The coup was designed to overthrow the *freely elected* leader of the country, Jacobo Arbenz Guzman, and replace him with a hand-picked, right-wing military officer, Colonel Carlos Castillo Armas. According to the story, the publisher of the *Times* did not know of the planned coup. Think of the implications of this event: Arthur Hays Sulzberger, the publisher of the *New York Times,* worked covertly with Allen Dulles, the director of Central Intelligence, and even wrote to Dulles about the reporter in question, "Of course, now that we have been alerted I shall watch his work with a great deal more care than usual."

Sulzberger put a beacon of the so-called free press in the service of Allen Dulles, one of the architects of the Cold War and founders of the National Security State, which threatens democracy to this day. Many books detail Dulles's role in bringing Nazi rocket scientists, Nazi psychiatrists, and Nazi intelligence agents to work for the CIA. During his tenure at the CIA, the United States government aligned itself with the forces of international fascism on every continent. This was the CIA that was intimately involved in the investigation of UFOs.

The *Times* article continues, "Mr. Sulzberger's files and a newly declassified C.I.A. history of the Guatemala coup . . . include more details of Mr. Sulzberger's cold war contacts with Mr. Dulles than previously published accounts." The reporter on this story observed that "Contacts of this sort between the C.I.A. and the American news media—as well as far deeper relationships—were common in the 1950s and 1960s. . . ."

Here is a clear example of the mechanism of the Mindshift Hypothesis at work in 1954, a critical period in the history of the UFO phenomenon in America and in the coverage of UFOs by major media outlets such as the *New York Times*. How would the publisher of the *New York Times* have responded to a request from the director of Central Intelligence to slant coverage of UFO information, then classified at a level higher than the hydrogen bomb, or to not cover it at all, because of national security issues? Cooperation seems assured.

How early in the modern era of UFOs did the *New York Times* show evidence of biased coverage based on an obvious editorial policy?

"I went through the UFO stories in the *New York Times* strictly chronologically," said Peter Robbins, co-author of *Left at East Gate: A First-Hand Account of the Bentwaters-Woodbridge UFO Incident, Its Cover-Up and Investigation*. "I was researching my book, and found that the skewing of the stories started right at the very beginning in the summer of 1947. I spent quite a bit of time doing research at the main branch of the New York Public Library in Manhattan. I wanted to see where the skewing and the sarcasm came in, and when the articles reflected more attitude than information.

"So the first thing that I did was to go through the *New York Times* newspaper morgue. I began with the summer of 1947 and continued into the early 1990s. I looked for any article, whether it was a feature or a small piece, which carried the words UFO, spaceships, or flying saucers. I pulled up about one hundred articles over a period of nearly fifty years."

A review of the *New York Times* index shows that a story appeared on July 4, 1947: "USAAF drops inquiry into reports of strange objects flying at 1,200 mph over Western U.S.; denies saucers are aerial missiles; possible meteorological explanations." On July 6, 1947, on the front page, the paper reported, "Scientists mystified as reports increase; saucers reported seen over entire U.S." The stories from that day attributed the saucer sightings to "public imagination." In addition, the *Times* reported that David Lilienthal of the Atomic Energy Commission denied that flying saucers were linked to atomic research. The *Times* also ran an editorial in that issue, "Those Flying Saucers." There were front-page flying saucer stories on July 7, 8, and 9 of 1947, in which the paper noted the "humorous skepticism of scientists and military experts" and that the "latest flock of rumors showed increasing imagination."

Although military experts may have expressed humorous skepticism to the *Times*, another story from that three-day period informed readers that military planes were searching the skies over the West Coast for flying saucers and that new reports of flying saucers were coming in from as far away as Sydney, Australia. The *Times* also published a piece on the views of a college professor who claimed he could prove that flying saucers were only "visual images."

On July 9, 1947, the Times reported that a flying saucer was seen near a New Mexico atom bomb site but that "it was found to be a weather balloon." With this first mention, the Roswell story had begun. On that same day, there were also reports of flying saucers from Great Britain and Australia. The subject of flying saucers was taken up at the United Nations because reports had aroused world interest.

On July 10, the *New York Times* interviewed a Professor Crespi who dismissed the whole thing as "probable mass delusion." On July 11, the *Times* asserted that public interest in flying saucers was lagging, even though it noted that saucer reports just kept coming in. By July 12, the *Times* was reporting on fake flying saucers that were being used by advertising agents in publicity stunts.

On August 10, 1947, a story in the *Times* carried a USAAF denial that a small boat had been destroyed by flying saucers off Tacoma, Washington. The next month, on September 28, an article appeared about flying saucers sighted over the New York metropolitan area. According to the *Times*, the witnesses were actually seeing "balloons used in a cosmic ray study." Finally, on December 27 of that year, the *Times* reported the views of a Dr. C.C. Wylie who linked flying saucers with "mass hysteria."

In retrospect, the Roswell incident is seen not to be an isolated event. The press coverage documents the flying saucer build-up from Kenneth Arnold's sighting on June 24, 1947, through the events at Roswell in early July 1947. From its first reporting on Roswell, it is clear that the *New York Times* has completely accepted the government's story uncritically. A bemused, cynical tone is apparent in the rest of the reports.

"The UFO phenomenon began with the unfortunate phrase, flying saucers," Peter Robbins commented. "UPI, Associated Press, Reuters all had reporters at local airports at the time of the Arnold sighting, waiting to pick up any information on the downed aircraft Arnold and the others were searching for that day. When Arnold landed, he told reporters what he had seen— a number of unidentified objects flying a great speed. When reporters asked what they looked like, Arnold said they flew like saucers skimming across the water. The idea of a flying saucer was seized upon in the imaginations of the reporters covering the story. It appears, from reading these old newspaper stories, that they almost could not help but put a certain sarcasm in their reports because of the term flying saucers.

"What we see in that first sweep of articles," Robbins said, "starting in the beginning of July, picking up with the Roswell incident, and then growing throughout the summer, as reports came in from around the U.S., Europe, and other international locations, was that the news coverage began with something that seemed humorous. From the beginning, the reports were almost all devoid of seriousness."

Robbins pointed out that, inevitably, the *Times* would call on the opinion of an expert, usually someone with a technological or a mental health background. Quite a number of psychiatrists, psychologists, and engineers were asked to make statements addressing the observations of regular people. From his review of the news stories, Robbins saw that "The remarks of these experts were almost always sarcastic and mocking. One of the more memorable pieces in the *New York Times* ends with a quote from Lewis Carroll's *Alice in Wonderland*. The implications were obvious—this is pretty goofy stuff."

According to Robbins, as the years went by, a pattern of reporting began to institutionalize itself. The reports would come in. The *Times* and other papers would cover them fairly dutifully. In the initial article, or in a follow-up piece—and there was almost always a follow-up—a person with a technological, mental health, or military background would be quoted. Robbins pointed out that "One of early military figures to appear in these stories was Curtis LeMay, who later became a ranking general and head of the Strategic Air Command. He was a hardcore skeptic and debunker. The debunking message would always be that flying saucers were not worth looking into. LeMay was quoted regularly that summer, and he was one of the first major names to appear."

Robbins observed that it was in this period that the professional debunker emerged. Today, professional debunker Philip Klass is the heir to that fifty-year-old tradition. However, in the 1950s and 1960s, the man most often turned to by the *Times* and other papers as the debunker of choice was a Harvard professor of astronomy, Donald K. Menzel. Dr. Menzel, the so-called voice of rationalism, promoted the basic idea that any UFO sighting could be explained away in conventional terms. What seemed beyond the pale was simply illogical. It could not be; therefore, it was not. To Menzel, flying saucers reports were all based on illusion, misinterpretation, fabrication, or a delusion.

According to Robbins, "From the 1940s onward, serious UFO witnesses who would not back down from a given reportage were alluded to as being less than mentally stable. This pattern has maintained itself almost to the present. However, we have been so inundated with UFO information in the last few years that there are now people in the media who have begun to make an attempt to report this seriously. The best early reporting and editorializing on the possibility that there might be something to the UFO phenomenon almost invariably came from smaller newspapers around the United States, not from any of the big media outlets. It almost seems that fear of ridicule was institutionalized."

Menzel was called on time after time as the authoritative source in articles on UFOs in the *New York Times*. He was the voice used to explain away the sightings and tell the general reader of the *Times* that the UFO was merely the reflection of light off crystals in the ionosphere or some other natural, physical phenomenon. Robbins said, "This pattern of reporting—initial report, skepticism, sarcasm, official explanation—continued on fairly unabated until 1952. Then in July 1952, we had the legendary UFO sightings over Washington, D.C."

A few months earlier, in April 1952, three months before the flying saucers invaded the air space above Washington, D.C., the *New York Times* published an editorial, "The Saucers Are Back," in which the paper commented on an article that had recently appeared in *Life* magazine. The editorial called flying saucer accounts "fantastic" and labeled so-called believers "romantics." Whereas *Life* seemed to take the Arnold sighting, and a July 1948 UFO sighting by an Eastern Airlines crew seriously, the *Times* dismissed all flying saucer reports as weather balloons or the planet Venus, or the results of self-hypnosis or attacks of vertigo.

In terms of the Mindshift Hypothesis, it is worth noting the contradictory nature of these mainstream media reports. They are just enough to get the public thinking about the subject, but they are not threatening. They appear to be examples of the creative dissonance that Michael Lindemann noted would be part of an effective public UFO education campaign. Are these stories the result of information planted by the clandestine group Dr. Mitchell says existed, the MJ-12 of Stanton Friedman, or the control group Philip Corso claims acted behind the scenes to simultaneously reveal and conceal the truth about UFOs?

Whether they were or were not, in July 1952, reality picked up where the 1951 film *The Day the Earth Stood Still* left off. In the movie, a flying saucer lands in broad daylight in Washington, D.C. On July 19, 1952, just before midnight, a flotilla of spaceships flew over the nation's capital, penetrating highly restricted air space. Over the weekend of July 19–20, there were numerous sightings in and around the Washington area. The "Invasion of Washington" continued one week later, on July 26 and into July 27. Skeptics had frequently complained, "Why don't these flying saucers just land on the White House lawn?" For two weekends in a row, UFOs made themselves visible to the American nation and the world in the skies over Washington, D.C.

According to Robbins, "There are some remarkable photographs of V-formations of lights over the Capitol. They were absolutely real. There were many ground-to-air observations, air-to-air observations, and radar observations. Strangely, Air Force jets were not scrambled immediately. This caused quite a stir at the time. Imagine the kind of priority this event would be given by the military. The official responses at the time were ludicrous and just didn't wash."

In the days before the Washington sightings, flying saucers were quite active. On July 14, the *Times* reported on an Air Force probe of a sighting of eight discs flying in formation 150 miles south of Washington, D.C. On July 17, Pan American pilots saw flying saucers moving at one thousand miles per hour and witnessed them climb one to two thousand feet in just seconds. On July 22, the paper reported that the U.S. Air Force had picked up unidentified objects, "perhaps saucers," on radar over Washington. This was the first acknowledged radar sighting of flying saucers. Yet no military jets were sent to investigate. In addition, commercial airline pilots also reported objects in the Washington vicinity at that time. Captain Casey Pierman said, "They were like falling stars without tails." At that time, New York City residents reported seeing similar objects over Central Park. On July 28, the *Times* ran a story in which the Air Force said that its jet interceptors had been outstripped by four to twelve objects that had been spotted over Washington for the second time in a week. Radar indicated that the objects were indeed solid.

In a *Times* story on July 29, Air Force officials insisted they were prepared to face the challenge of UFOs and tried to explain their twenty-four-hour lag in chasing the mysterious objects that were flying over the center of government. A *Times* editorial that same day wrote of "the romanticists who are willing to believe that Mars and Venus are inhabited by people so intelligent and advanced that they have been visiting us for centuries." On July 30, the Air Force issued a public statement asserting that the objects it had been chasing with jet aircraft were natural weather phenomena. The *Times* ran a page-one story, "Air Force Debunks Saucers As Just Natural Phenomena." A military official was quoted as saying, "None of the several thousand saucer reports checked by the Air Force in the last six years has disclosed the existence of [UFOs]." The following day, a *Times* editorial supported the official U.S. Air Force policy.

On August 1, a story appeared in the paper stating that the Pentagon had received 432 flying saucer reports. One week after the Air Force debunking story, the *Times* reported, on August 7, that unknown objects were again spotted over Washington, picked up on radar, and that, again, Air Force jets were no match for the unknowns.

At the time, the newspaper of record attempted to explain away the sightings as fantasies, comparable to such things as the Loch Ness monster, or as misinterpretations of physical phenomena such as "radar ghosts" or temperature inversions.

Reports continued in this general debunking pattern through the 1950s, including the 1957 American UFO Wave, a series of UFO sightings that surpassed even the 1952 wave. The night of November 2–3, 1957 is, by some accounts, the single greatest night in UFO history, a night on which there were sightings that ringed the globe. Yet this wave received little coverage in the *Times*. There were, however, other interesting reports in the newspaper of record in 1957. For example, on January 17, 1957, retired Rear Admiral Fahrney told the *Times* that high-speed, directed objects are entering Earth's atmosphere. The Air Force replied that it had no concrete evidence to support the Admiral's claim. And on November 6, 1957, a Bell Aircraft engineer was quoted in a story as saying that the reported unidentified objects came from outer space.

On January 24, 1958, the *Times* ran a story on an unusual form of censorship involving the CBS television network. Donald Keyhoe, author of *Flying Saucers Are Real*, was participating in a television discussion of flying saucers when he was cut off as he was about to make comments about UFOs that the government had not previously approved. CBS went along with this censorship and turned off Keyhoe's microphone as he spoke. He intended to tell the viewers that UFOs would be proved to be real if Congressional hearings could only be held on the subject.

The *New York Times* did not stand alone in its derisive treatment of the UFO phenomenon. The major U.S. magazines of the day also reflected the editorial policy evident in the *Times*. During the 1950s, they published articles with titles such as:

**Flying Saucer Hoax** (*Saturday Review*, December, 1952)
**Flying Saucers: New in Name Only** (*Readers Digest*, August 1952)
**Saucer Season** (*Newsweek*, August 11, 1952)
**Those Flying Saucers** (*Time*, June 9, 1952)
**The Great Saucer Hunt** (*Saturday Review*, August 6, 1955)
**Waiting for the Little Men** (*Newsweek*, March 28, 1955)
**Cups or Saucers?** (*Time*, September 9, 1957)
**Seeing Things** (*Newsweek*, November 11, 1957)

The same pattern continued into the 1960s. In 1966, the *Times* carried three reports of UFOs landing near Ann Arbor, Michigan. About forty witnesses saw four "sister" ships set down near a swamp. The debunking cover story released to explain away the sightings—that it was all just swamp gas—caused a national uproar, and great scorn and ridicule were directed at the government's preposterous story.

Dr. J. Allen Hynek, an Air Force advisor on UFOs and the head of the astronomy department at Northwestern University at the time, was sent to investigate the story. However, he was hung out to dry by the government officials who ran for cover in the face of the intense public backlash to the absurd swamp gas explanation. In September 1966, Hynek called for a scientific study of UFOs.

Hynek's suggestion led to the Condon Committee and the biased Condon Report. Condon himself had already prejudged the matter at hand and simply went through the motions. His contemptuous attitude toward UFOs was frequently demonstrated in public and led to dissension on the staff. Important investigators quit in disgust. The whole study was a fiasco and a debacle.

In 1968, the Condon Report was released, finding no evidence that UFOs are intelligently guided spacecraft and concluding that any further study of the subject was a waste of taxpayer's money. The *New York Times* accepted the Condon Report at face value. The paper did not look at the data in the report. A careful review of Condon's own data suggests that a large percentage of the report's own UFOs cases remained unsolved. Be that as it may, the U.S. Air Force used the report to justify closing down Project Blue Book, bringing an *official* end its tortured involvement with UFOs.

The *Times* followed the reactions to the controversial report throughout the year, concluding with an article on December 28, 1969. According to the report, astronomers, physicists, and

*Art imitates life as the film* Unidentified Flying Objects *incorporates actual an 1952 UFO event into its plot.*

social scientists convened for the annual meeting of the American Academy for the Advancement of Science, called on the Air Force to preserve the secret UFO records. The scientists feared that the military would use the Condon Report as an excuse to destroy all the records of the entire UFO period.

The next major UFO wave took place in 1973, when most American newspapers were focused on the unfolding saga of Watergate. On September 15, a month before the October peak in the wave of sightings, the *Times* reported that a little-known Southern governor—James Earl Carter of Georgia—told the press that he had seen a UFO before he became governor. On October 15, in response to sightings around the U.S. and the world, the U.S. Air Force and establishment astronomer-debunkers told the *New York Times* that they were not taking reports of UFOs seriously.

On October 17, the *Times* informed its readers that UFO reports were coming in from various parts of the U.S. and that there were sightings from the Soviet Union. The paper also reported on a network of four stations that appeared to be receiving radio signals in the form of pulses that they believed may be coming from another civilization. The pulses were received regularly, several times a day, and lasted for several minutes each time. Soviet scientists ruled out artificial satellites or solar flares as the source of the pulses. In addition to the nationwide sightings, a meteorologist in Mississippi claimed that his radar equipment failed as he was successfully tracking a UFO. On October 18, the *Times* reported that Governor Gilligan of Michigan and his wife saw a UFO near Ann Arbor on October 17.

True to form, *Newsweek* ran a piece entitled "Stardust and Moonshine: UFO Sightings," on October 29, 1973, linking UFO sightings with drunkenness and hallucinations in the reader's mind. On November 3,1973, *U.S. News & World Report* published an article called "Are Flying Saucers Real?" Interestingly, the *Times* provided no coverage of the widely reported alien abduction of Charles Hickson and Calvin Parker that occurred on October 11, 1973, in Pascagoula, Mississippi. But the mocking, debunking tone was still prevalent.

According to the results of a Gallup Poll reported in the *Times* on November 29, 1973, twenty-six years into the modern era of UFOs, 51 percent of the American public believed UFOs were real; 11 percent claimed to have seen a UFO; and 95 percent had heard or read about UFOs.

Peter Robbins studied a fascinating example of UFO coverage in the *New York Times*. In 1989, a series of articles captured the essence of the editorial policy of the paper in marvelous detail and clarity. Robbins recalled that "In October 1989, there was a Russian UFO incident in Voronozek, a city about four hundred miles south of Moscow, I believe. There was a landing *and a disembarkation.* All kinds of physical evidence was available—physical impressions on the ground, beta and gamma background radiation readings—as well as a huge multiplicity of eyewitness accounts by youngsters, seniors, and highly technically trained people.

"I remember when the report first came in. I heard it on the radio. I was living in Manhattan on 57th Street at the time and I called Tass, the Soviet news agency, which still maintained an office then in Rockefeller Center. I explained who I was and asked for *original teletypes* of the material. They were extremely forthcoming and provided me with the original material. They took it deadly seriously. This UFO incident set up one of the most interesting patterns of reportage I have ever seen in the *New York Times*.

"The very first day, it was front-page news. This has happened before, but not very often, at the *Times*. It was quite a serious article. The second day, it was either front-page or page two. But it was now a totally mocking, gooney, silly article. On the third day, the *Times* published a really weird, mixed kind of article. There was a day off and then there was a comeback, which was once again quite grounded. There then followed an editorial cartoon and, that weekend, a full editorial on the so-called Russian sighting.

"By the end of the week, however, the *Times* had decided to put forward a story claiming to show that, with the disintegration of the Soviet Union in the autumn of 1989, a number of unusual things were occurring as communism was dying. One of them was that, as we all know, underneath the Marxist regime, the Russian people are deeply mystical. Now all of these mystical longings were starting to come forward. The Soviet press was, for the first time, experimenting with tabloid-style treatment of this material, and so there were now articles on ghosts and UFOs and paranormal stuff.

"It is a particularly remarkable series of articles. At its end, a *Times* editorial discussed Dr. Robert Goddard, the father of rocketry in America. In the early 1920s, Goddard predicted that, in the not-too-distant future, man would be able to send a rocket to the Moon. In his day, Goddard was laughed at. The last sentence of the *Times* editorial is one of the most remarkable pieces of Orwellian double-speak that I have ever seen in my life."

According to Robbins, the editorial linked Goddard's vision—which was laughed at, but which later became a reality—with Tass's stories of UFO landings in Russia. "I think the sentence read something like, 'As surely as rockets will not fly in space, the story of the century happened in Russia this week,' or 'As surely as rockets have never flown to the moon, Tass has broken the story of the century.' However it was phrased, it was an infuriating quote. It actually scared me. Orwell pointed toward this dark direction of giving non-information. This editorial comment is Orwellian double-speak. It almost sounds really important. Yet it is just a smug bit of absolute nonsense.

"In this series of stories, we had the basic pattern that has informed UFO reporting in modern times. One of the results of this reporting has been that, both the strong and the weak, the powerful and the powerless, have all been magnificently conditioned over five decades to wire up an even remotely serious report of UFOs with mental instability, hallucinations, and insanity. People are still reticent to come forward with a serious opinion on this subject in a public forum."

In the mid-1990s, UFO coverage in the *Times* was like television reruns. The newspaper dragged out all the hoary old comparisons: UFOs and the Loch Ness monster; UFO sightings and mental illness; and UFO interest and the public's "mania for monsters." In 1947, the *Times* uncritically accepted the Air Force's weather balloon explanation for the Roswell crash and, in 1994, the paper uncritically accepted the Air Force's "new improved" weather balloon explanation for Roswell. On December 11, 1994, the *Times* mocked support groups for abductees. In 1996, news stories linked the Roswell incident to Hollywood movies such as *Independence Day*. And in 1997, the *Times* ran many debunking articles related to the fiftieth anniversary of the Roswell crash. One such article appeared on June 15, 1997, "Indeed They Have Invaded. Look Around." The *Times* portrayed flying saucers as "the center of a modern mythology." As is its wont, the paper compared UFOs to unicorns and leprechauns.

On June 25, 1997, the *Times* ran a very large page-one photograph of the Air Force's crash test dummies themselves. A full-page story, "Air Force Details a New Theory in U.F.O. Case," inside featured Air Force photographs of a saucer-like mock space probe at the White Sands missile base; a test dummy insulation bag "that may have been mistaken" for a body bag; and a crash-test dummy with Roswell eyewitness descriptions of alien bodies keyed to the features of the dummies. The paper accepted, with a completely straight face, the even-newer and more improved Air Force explanation for Roswell—that crash-test dummies dropped from the sky in 1953 were seen in the New Mexico desert in 1947. Now, there's a story!

The major magazines joined the *Times* in the Roswell debunking fun. *U.S. News & World Report*, *Newsweek*, *Time*, the *New Yorker*, and many other national publications spoofed any serious interest in Roswell in particular and UFOs in general. The prevailing mindset was entrenched and would not be moved one inch.

The Mindshift Hypothesis puts forth the assumption that the press may be used to re-educate the American public concerning extraterrestrial life. Peter Robbins has studied the manner in which the press in general, and the *New York Times* in particular, has covered the UFO phenomenon. His opinion on the matter is worth considering.

"I would say that the Mindshift Hypothesis is certainly within the realm of possibility or even probability," Robbins said. "For decades, there has been what amounts to a campaign of debunking. This was what most informed serious reporting on the UFO phenomenon. It would not surprise me, based on my studies, if there were a clandestine group influencing the major media outlets. Let's start with Dr. Menzel. He was, in fact, a distinguished professor of astronomy at Harvard and had a long and distinguished career as an academic. He wrote several books on astronomy. He also was the author and co-author of several books on UFOs. Literally, everything in Menzel's UFO books is about explaining away the phenomenon. What we now know about Dr. Menzel makes him a very good case in point.

"Stanton Friedman is the first researcher to have the opportunity and the permission to review Menzel's papers at Harvard; to review all of his private papers with the permission of Menzel's widow; and to go through the papers at the National Archives. Through his diligent research, we now know—and this is not speculation—that Dr. Menzel lived a double life. He was a consultant to the U.S. government in the original national security council, even before it was called the National Security Council. He was an insider's insider, but this was not in any way public knowledge. The world that knew of Menzel knew of him as an academic associated with one of the most distinguished universities in the United States. I refer you to Friedman's work on this for detail. His original lecture on this topic at American University in the summer of 1987 was one of the best lectures I have ever heard."

In 1947, it was not known if UFOs might not turn out to be a sinister reality. If this proved true, then there was potential for social destabilization and mass panic. There was, at the time, grounds for responsible leaders to take that possibility seriously.

Robbins observed, "If that was the case, then let's face it, all we would need is one discreet, buttonholing incident between Sulzberger, who was then publisher of the *Times*, and somebody high up in government to say, 'Art, we need to enroll you to set a tone on this, so that while we're trying to figure out what's going on, nobody will see the major respected organs of the media, as exemplified by the *Times*, taking this UFO phenomenon seriously.'

"But I would also have to say, in all fairness, that even if there were never any contact between media and government, never a wink and a nudge, or a sincere request by the government for the official attitude to be reflected in the media, the powers that be at the *Times*, or in the Western press in general, may simply have taken it upon themselves to do so until the government came forward with something serious on the subject. On their own, they may have said, 'It is our duty to belittle, debunk, put down, or demean this phenomenon.'"

Today, more and more excellent work on the UFO phenomenon is seeing the light of day. Slowly, people are becoming increasingly open to this knowledge. "Every time someone like a Timothy Good, or a Budd Hopkins, or a Dr. David Jacobs produces a solid serious volume, and continues to build a serious readership," Robbins said, "the paradigm again shifts, one person at a time, and moves forward."

Peter Robbins is contributing his share to the advancement of this awareness as well. And it seems he is getting some support along the way. "On January 28, 1998," he said happily, "Dan Ackroyd was a guest on Rosie O'Donnell's show. By the way, she takes the subject of UFOs seriously indeed. They finished talking about his new Blues Brothers movie and she asked, 'What's new on the UFO scene?' At which point, Ackroyd reached into his pocket and pulled out a small promotional copy of my book, *Left at East Gate*. He said it was an important book and recommended that she read it." On that show, many people were introduced to a new book on the UFO phenomenon and, as people read it, we will move one more step forward in our understanding of the UFO enigma.

On September 24, 1952, H. Marshall Chadwell, the Assistant Director for Scientific Intelligence, wrote a memo to CIA Director Walter Smith, about flying saucers, in which he recommended, "to minimize risk of panic, a national policy should be established *as to what should be told the public* regarding the phenomena." There is some evidence, according to debunker Peebles, writing in his book *Watch the Skies!*, that Chadwell believed that the Extraterrestrial Hypothesis was a real possibility.

Eventually, the CIA formed a panel of experts to evaluate the UFO phenomenon. The Robertson Panel opened its first meeting on January 14, 1953. The "Report of the Scientific Advisory Panel on Unidentified Flying Objects," prepared by the Robertson Panel, contained a memorandum written to the assistant director for scientific intelligence of the CIA on January 17, 1953. The panel recommended, in section 4, paragraph (b), that "the national security agencies institute policies on intelligence, training and *public education* [author's emphasis] . . . an integrated program designed to reassure the public of the total lack of evidence of inimical forces behind the phenomena. . . ."

There is ample evidence that the U.S. government has long used the mass media to advance its own purposes, such as preparing the public for war, concealing the development of the atom bomb, or promoting the civilian use of atomic power. Why wouldn't the clandestine group within government use the major print media outlets to advance its own agenda regarding UFOs and extraterrestrial life? And why wouldn't the corporate entities and media moguls participate and play their part in the public education campaign recommended by the CIA's Robertson Panel?

On the 23rd Aug. 1974 at 9 o'clock I saw a UFO. —J.L.

*from the back cover of John Lennon's* Walls & Bridges

# CHAPTER NINE

# Mindshift in Music
# Mindshift in Music
# Mindshift in Music
# Mindshift in Music
# Mindshift in Music

A fascinating connection between music and the UFO phenomenon was made in the 1977 film, *Close Encounters of the Third Kind*. In the movie, the extraterrestrials and human beings communicate at first through a simple melody. At the climax of the film, when the Mother Ship appears over Devil's Tower, the musical communication becomes so complex it can only be handled by human computers interacting with the alien spacecraft.

Spielberg included this essential element of the story in the film long before author Gerald Hawkins linked crop circles, which are frequently associated with UFOs, with the ratios of the diatonic musical scale in a 1992 article published in *Science News*.

However, for most Americans, the link between music and flying saucers was made on the radio. Rock 'n' roll was born between the two great American UFO Waves of 1952 and 1957. Although the rock 'n' roll songs of the 1950s were almost entirely about teenage love and sexuality, there were a number of songs about flying saucers and spacemen.

In 1956, "Flying Saucers," by Buchanan and Goodman, went over the radio into American homes and out to the Universe. The first rock 'n' roll song about UFOs to fly high into the Top 40 charts was a humorous novelty single, called "Purple People Eater" by Sheb Wooley, released in 1958. This song reached a mass audience through Top 40 radio airplay and is the first pop tune about saucers ever heard by many of the emerging baby boomers. That same year Billy Lee Riley and the Little Green Men had a hit with the rockabilly tune, "Flying Saucer Rock and Roll."

Two comedians released comedy cuts about saucers and outer space: Bill Dana in 1962 with "The Astronaut," and Alan Sherman in 1963 with "Eight Foot Two, Eyes of Blue," which appeared on his album "My Son, the Nut." In 1963, the Top 40 hit "Telstar" celebrated humanity's space satellite of the same name. The Ran Dells released a song called "Martian Hop" in 1963, and the Marketts had a hit in 1964 with "Outer Limits."

Rock 'n' roll songs about flying saucers in the 1950s and early 1960s conveyed a pop version of the standard UFO message that flying saucers were not to be taken seriously. Saucers were in a category not unlike other fads, such as Hoola Hoops, swallowing goldfish, and stuffing college students into phone booths. Flying saucers were just good, clean fun.

In the early 1960s, the immensely popular and influential young folk singer, Joan Baez, took America by storm. In the music industry, she combined purity of soul with purity of voice and brought conscience to commerce. On her album, *Joan Baez Vol. 2*, she performed a moody and mysterious ancient ballad called "Silkie." This song tells the story of the Great Silkie of Sule Skerry, a being who lives in the depths of the earth, below the oceans. The Silkie mates with an earthly woman, who bears a child. The Silkie later returns to claim his son, a hybrid of the two beings. But when he does, he also foretells the death of both his son and himself at the hands of human beings.

It is a fascinating song about the extinction of a friendly race of intelligent nonhuman beings who once interacted intimately with earth people. In this beautiful and haunting ancient song, many themes of both folklore and the UFO phenomenon are present. Because of Joan Baez, it was heard by a wide audience in America and beyond.

A dramatic change occurred in the United States in the years following the assassination of President Kennedy. Societal pressures that had been building up for decades came to a head. In many ways, this shift was most obvious in what became known as the youth culture. The attitude of the new generation toward authority was, in many ways, antithetical to that of the generation of World War II and the Cold War era. This "generation gap," as it was then called, was evident in family life, in education, in religion, in politics, and in the entertainment industry as well.

In 1963, the Beatles reflected this new attitude dramatically as they shattered the Tin Pan Alley mold that ruled the pop music industry. They wrote their own songs, arranged them with a producer's help, and performed them as well. Before the Beatles, most pop groups performed songs that were written and arranged by professional tunesmiths, chosen for the group by a producer, and recorded by professional engineers and producers assigned by the record company. After the Beatles, an entire generation of singer-songwriters emerged and produced material on a wide range of topics, far beyond the packaged sound of Tin Pan Alley pop music. This change in the industry allowed songs about UFOs and extraterrestrial life to reach hundreds of millions of minds.

In 1966, the groundbreaking "folk-rock" group, the Byrds, released their song "Hey, Mr. Spaceman." The song hit the charts in the same year that the Barney and Betty Hill abduction story was splashed across the pages of *Life* magazine and reached the best-seller list in John Fuller's classic, *The Interrupted Journey*. The lyrics tell the story of what we would now call an alien abduction. The singer woke up to see those "saucer shaped lights that put people uptight." But instead of being frightened, he pleads, "Hey, Mr. Spaceman, won't you please take me along?" The song concludes with a message from the spacemen that many abductees report hearing, "So long, we'll see you again." This is a jaunty, upbeat song with a message that is very different from earlier pop songs about UFOs.

The San Francisco bands, the Jefferson Airplane and the Grateful Dead, were the founders of "psychedelic music" or "acid rock." At first, the Jefferson Airplane wrote songs that fit the commercial radio format of the day, and they had a huge hit with "Somebody to Love." After their second hit song, "White Rabbit," they became increasingly radical and complex. Soon their music was heard only on FM radio stations with a "progressive rock" format.

In 1970, the group released a single entitled, "Have You Seen the Saucers?" In the song, the Airplane ask, "Do you know there are people out there who are unhappy with the way that we care for the Earth Mother?" Referring to the U.S. space program, the lyrics are strong and biting: "Tranquility Base—there goes the neighborhood, an American garbage dump in space." In the refrain, the song posed a question to the audience, the "first born of the atomic generation," about the U.S. government and UFOs, "Have you any idea why they lie to you?" The messages of the song—we are damaging the planet; we are not alone; and the government is covering up UFO and ET presence—resonate strongly today, twenty-eight years after its release.

In late 1970, Jefferson Airplane underwent a mindshift and "morphed" into Jefferson Starship. One side of their new release, *Blows Against the Empire*, was composed of a series of songs that told the story of "people with a clever plan who would assume the role of the mighty" and hijack a starship that was being built in Earth orbit. The bold rebels that Starship sang about planned to seize the starship and sail off into the Universe to build a better world. The basic assumption of the musical piece is that the Universe is populated and that humanity can find a place there, a concept that is more widely accepted today.

A current rock icon also released a UFO-related song in 1970. Neil Young's "After the Gold Rush" appeared that year on the album of the same name. (In 1996, Linda Rondstadt performed the song on her latest album.) In surreal lyrics that carry the listener from a dream of knights in armor to a burned-out basement, the singer observes that humanity has "Mother Nature on the run in the 1970s." In the last verse of the song, the singer dreams that he sees silver spaceships in the yellow haze of the sun. In his dream, he sees that "the loading has begun." The "chosen ones" are now "flying Mother Nature's silver seed to a new home in the sun." UFOs are in some way a phenomenon of Nature.

Again, the theme of human beings destroying Nature emerges and is linked with the arrival of UFOs. However, in this song, the spaceships are "Mother Nature's silver seed" and are taking only a chosen few humans to a new home. Many abductees report being shown natural catastrophes that decimate humanity. Some seem to feel that aspects of alien abduction activity may involve saving specimens of living things for later use, as a kind of cosmic "Noah's Ark." This song is quite prescient, containing UFO information that did not become widely known until almost twenty years later.

In 1974, in the liner notes to his album, *Walls and Bridges*, John Lennon wrote that he had seen a UFO over New York City. Ten years later, in his song "Nobody Told Me," (which appeared on the 1984 *Milk and Honey* album released after Lennon's murder), Lennon sang, "There's UFOs over New York, and I ain't too surprised."

PolyGram Records, 1983, 1984.

The popular comedy group, the Firesign Theater, released a hilarious spoof on UFOs and many other "New Age" subjects in the album, *Everything You Know Is Wrong*. They satirized the kooky side of the UFO phenomenon: the world of confused, misguided "seekers" who fall prey to the con artists, hoaxers, and frauds who exploit the interest of the naive and gullible. The motto of the UFO expert on the album is, "Don't feel alone out there in the New Age—there's a seeker born every minute." As the Firesign Theater poked fun at the follies and foibles of "believers," it delivered an important message, "Everything You Know Is Wrong; The Brain Is Not The Boss!" When facing an intimidating unknown phenomenon, humor can bring relief and a healthy sense of honesty about what is known and not known about UFOs.

British ufologist Jenny Randles has written about the December 1980 Bentwaters-Woodbridge UFO incident that occurred at an American nuclear base in the United Kingdom. In

*Left at East Gate*, Peter Robbins has also written about Bentwaters. In his excellent book, Robbins notes that in 1984, singer-songwriter Chrissie Hynde, of the rock group the Pretenders, wrote a song about the UFO incident called "Show Me." The song, which became a top-ten hit in England and which was also very popular in America, is addressed to the aliens from outer space, welcoming them to the human race, "with our wars, our greed and insanity." The song brought one of the most important UFO incidents of our time to millions of listeners.

Many other groups and performers, for example, Blue Oyster Cult, David Bowie, Foo Fighters, and UFO have also produced songs about UFOs over the years. However, the commercially successful performers and groups are only the tip of the iceberg. The mindshift about UFOs and extraterrestrial life that popular music is helping bring about occurs every day of the year in performances by talented individuals who do not have—or yet have—recording contracts from major studios. They play in clubs in every corner of the country and reach a new generation open to the reality of UFOs and ET presence.

Both UFO songs and UFO experiences have shaped the lives and careers of two contemporary women singer-songwriters who are representative of a great many performers across the country known only to a relatively small number of fans. Their songs about UFOs and their personal interactions with ETs are highlights of their live performances. Some UFO-related tunes are included on their CDs.

Jane Allyson is a singer-songwriter based in New York City. In 1979, after a profound UFO sighting and series of encounters, she left music and began to practice as a healer. In 1997, she returned to her music professionally and began working on a CD—*Shanghai Lily Dublin*—that will be out in 1998. Her first album was already influenced by her interest in extraterrestrials, although she had

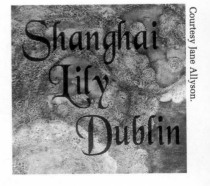

not yet had any experiences with UFOs or ETs. "My songs were romanticized versions of things that hadn't happened yet," she explained. "In 1979, I had my first UFO sighting and then things changed."

At the time she lived, rehearsed, and recorded her music at a studio in lower Manhattan. Throughout the time she lived and worked in that building, she said she felt an impulse to go up to the roof to see something. She didn't know what she would see. The impulse scared her somewhat, and she did not yield to it. However, one night, she did go to the roof. "I had to go," she said. "The first time I went, I didn't see a proverbial UFO. But I did see an arc of flashing lights over the Lower East Side. I went back in the daytime to see if the lights were simply part of a large sign of some kind. There was no arc-shape visible anywhere in the neighborhood. But on July 4, 1979, I went up on the roof with two friends to watch the fireworks. To the west, over the Hudson River, I saw six round lights in pairs of three, blinking on and off. Suddenly, they were over the East River."

At first, Jane thought they were helicopters. She watched them drift slowly for awhile, thinking they might be involved somehow in taking pictures of the fireworks display. The lights moved over toward the Empire State Building and seemed to be, not helicopters, but a huge object. "I was mesmerized," she said. "Then I thought, 'Come closer!' And then there it was, overhead. It wasn't six separate flashing lights at all but a diamond-shaped object the size of a six-story building. I didn't see any windows. The whole thing glowed like a very white light and had smoke or fog coming off the sides of it. The strange thing I remember about the object was that *it seemed both solid and non-solid.* I felt I could have put my arm right through it."

She noted that she and her companions each had different emotional reactions to the sighting. "I felt tremendous fear and worried that the object might crash down on us. I felt terror. My boyfriend became rigid and couldn't speak. He was almost catatonic. The woman friend who was with us was laughing and crying simultaneously." Jane's emotions began to shift. "I began to feel hostile. I wished I had a rifle so I could shoot the thing down—*now!* Although I felt hostile, I also felt wonder. I thought, 'If you don't leave now, I'll go crazy!' Then *it disappeared.*"

The three young people went down to their home. Jane wanted to report the sighting, but the other woman stopped her from doing so. She was startled to see her boyfriend meditating, something he had never done before. Her female friend began to dance about the room nervously. Jayne said, "I felt profound, unbelievable compassion for all of the human race, all people. I felt compassion flowing out of me like water." Her feelings persisted for awhile and then ceased. She called the police and the newspapers to report the sighting and was troubled that there was no corroboration by others.

Following the Fourth of July sighting, she experienced what she calls a "visitation." At 4 A.M., in bed with her boyfriend, she got up and went downstairs and lay on the couch. She felt paralyzed and terrified. She felt as if there were a heavy slab of cement on top of her. Next, she saw three small dark figures, between two and three feet tall. They wore uniforms that covered the entire body except the face. The beings communicated to her telepathically, "If we can do this to you, we can do this to your nuclear weapons." She awoke terrified and did not tell anyone about it at the time, not even her boyfriend.

Jane's second 1979 experience was quite different. She woke up in the middle of the night to see a small, charcoal-black UFO outside her window. A man in a blue uniform, a life-size human form, floated in through the window. "He was a beautiful man with tan skin," she said. "He took my hand and I floated out of bed and hovered about two feet off the ground. I thought to myself, 'I wonder if he's married?' He telepathically communicated, 'I do have someone where I come from.' I felt so much love for him and felt something from him that brought a tremendous diversity of love—for parents, sisters, cousins, friends. It was magnificent." She had never felt anything like this and, afterwards, wondered how she could settle for ordinary life. Jane watched as he "floated right through the heavily-gated window to the saucer, which shot away like a stone skipping on water."

Today, her music captures her feeling of ET energy. "My music comes from real experiences," she said. "Sometimes I feel other songwriters are just imagining how it would be." Jane's group is called Shanghai Lily Dublin, which is the name of the new CD as well. The group performs, and has recorded, three songs that are UFO-related: "Reflections of Your Ship," "Martian Love," and "Fall from Grace." She does all the vocals and arrangements and is working with producer Jeffrey Lesser on the CD. "Jeffrey Lesser won a Grammy for his work with the Chieftains," Jane noted. "He has also worked with Lou Reed, Elvis Costello, Sinead O'Connor, and Barbra Streisand."

Jane Allyson is performing her music again and no longer works as a healer. "In order to do my music, I had to leave the world. If you are *really* doing something, your whole life has to change. I threw a funeral for my old life and decided who I was going to be in the future. I have never gone back."

Drawing courtesy Helen Wheels.

*Portrait of Helen Wheels by R. Crumb*

Helen Wheels is a singer-songwriter now working with two different back-up bands, the Skeleton Crew and Work. She was the first artist to provide important UFO information, such as reprints of Freedom of Information Act UFO materials and a listing of reliable sources of UFO information, on an album cover.

One of the other members of Helen's band is—perhaps coincidentally, perhaps not—also an abductee. They write and perform UFO-related material, including a song whose title they cannot agree on. It is called both "The Saucer Song" and "The Truth Is Out There." Helen recently performed in New York's East Village at Brownies on Avenue A. The audience responded enthusiastically to "The Saucer Song." People came up to Helen and spoke to her about the song. "I do a rap in the middle of the song about the Freedom of Information Act and UFOs. People in the audience were saying, 'Yes! You've got to be political about this stuff!' I was just amazed at the number of people who said something about the song. Now all this is so much more out there. Fifteen years ago, it was why I felt so much like a freak. At this point in my life, it's just one of the things that has shaped me."

The UFO phenomenon has been a part of her life since childhood. Helen is an abductee and is completely open about it. She has worked with Budd Hopkins, the late Pete Mazzola, and other UFO researchers, and has spoken in public about her alien abductions. "Basically, they have snatched me up since I was a little kid," she said, "and this continued on throughout my reproductive life. When I was twelve, my brother, who was fourteen-and-a-half at the time, and I had a sighting together. He turned around to go in and tell my mother there was this flying saucer hanging over the neighbor's house when he was struck down with a blue ray that went over my shoulder. The ray came from the bottom of the center disc in a V-formation. He went down, kind of in slow motion. He was conscious at first, but eventually became unconscious."

The two siblings remembered the event for a couple of days, and they talked about it, but then the brother's memory of the event vanished. It did not return for many years until he was an art student. Then it returned all at once. Helen said, "I always remembered the sighting consciously, but I did not remember the alien abduction. I did see a Mother Ship at that time as well. I didn't remember the abduction aspect until my brother began to work in the UFO field. I met with an investigator, Pete Mazzola, who determined that this was a real case. He said it was very unusual for one sibling to be abducted and not the other. That's when I started considering doing hypnotic regression."

Helen had been plagued over the years by strange, disturbing dreams and decided to investigate the matter because it was less difficult to look into it than not to explore what may have happened. "I decided in the late 1970s that I would rather have information than just the fear. I worked with Pete Mazzola because he had cracked the case. But he was in a terrible accident and died. That's when I started seeing Budd Hopkins. I worked with him, on and off, for a long time."

In 1977, Helen performed a song about UFO abductions. She also performed other UFO-related songs at the time. "I had an EP called 'Post Modern Living,' a single called 'Room to Rage,' another single called 'Carry My Own Weight.' I also wrote songs recorded by other people such as Blue Oyster Cult." Helen had been interested in David Bowie's space songs and many of the other songs on that theme. "Just look at the history of rock 'n' roll. You see the classic 'Flying Purple People Eater,' 'Mr. Spaceman' by the Byrds, and the Airplane's song 'Have You Seen the Saucers?' Blue Oyster Cult did several songs on the subject, like 'ETI.' Patti Smith had one about Wilhelm Reich's son Peter and the saucers. There are quite a lot of other songs."

Although she has incorporated her abduction experiences into her music, Helen felt initially that the abduction experience was akin to rape, and it filled her with a vast amount of anger. "Creatures are doing stuff with you and your life is changed. I had a lot of anger. At the same time, it was like a magical initiation because at the age of twelve, I experienced telepathy. I had two types of creatures. One was a much bigger creature who basically tracked me through my childhood. And then there were the little 'dentist-office'-type horrible ones who do stuff to you—the Greys. The particular creature who tracked me through my childhood communicated with me through telepathy. He would tell me they were not going to hurt me, and then he would turn me over to the little creatures who did hurt me!"

Helen described the tall being as seeming to be male, although there was no sexual element involved as we know it. He was about six feet tall, ultra-thin, with a large head and black, shiny wrap-around eyes that looked like Italian sunglasses. "I know exactly what he was wearing. If I described it in modern terms, I'd say his clothing was like a blue, metallic Spandex. Like an airman's suit that goes right up to the throat and is skin-tight down to the wrist and ankles. It didn't have an insignia on it but it could have been a uniform. This being floated. He didn't walk."

It was during her regression sessions with Budd Hopkins that Helen became aware of the entity who had been following her all her life since childhood. She became aware that she always met with this particular being. Helen recalled, "He told me four main things repeatedly: we're not going to hurt you; we've seen you before; we'll see you again; and don't be afraid. Sometimes he would show me something fabulous. I have recollections of being in a control room in a saucer, with all the dials and switches. It was sort of like the Concorde. There was a star chair I could lay in. *And the stars don't blink.* He would let me look at all this stuff and give me all this programming that I wouldn't be hurt. And then I'd be in the next room—these rooms were shaped like a piece of pie—and there would be eight or nine of the little Greys. They would float me onto a medical table, put things up my nose, and do all kinds of painful things. They seemed to have no idea that they were hurting me."

When Helen thinks of the tall being, she feels that she was involved in something amazing that has changed her life.

However, her memories of the Greys simply leave her feeling hurt. She spent eight years in therapy, dealing with the anger and pain that resulted from the abductions. "I have a lot of curiosity but they don't let you ask the questions. They took genetic material from me and I'd like to ask them, 'What's up, guys?' They do these things without any kind of understanding of our emotions. They have no idea about emotions, is my take on it. I worked with Budd's support group in the city for many years. It was the initial place I went, and I still go back when I can. There aren't many places you can get help."

Her involvement with extraterrestrial intelligences has had a dramatic impact on Helen's religious views. "My religion has become Nature. I was raised Jewish but it has all changed for me. Nature is, for me, the place where I feel a connection with the Infinite or the Goddess. There is no temple or church or anything. I just need to be in Nature. At the same time, I feel a fierce protecting Spirit. I think that, basically, the aliens season you to be somebody that is ready to be the protector of Nature. Or perhaps this is a reaction to the experience. I look to Nature for help from the Universe."

These insights and experiences are reflected in Helen's music. Fifteen years ago, she was ahead of her time but today, things have caught up. "I think that Hollywood has totally normalized the UFO phenomenon for this generation. Kids now think, 'Sure there are ETs' or 'Sure there are spaceships.' It's not a freak-out. They brought the subject up in the United Nations twenty years ago—I was there. But they squashed it. They didn't even let it get out to the news that they were talking about UFOs at the UN. It seems different now and it's a lot easier. I'm just a searching artist who is compelled to create in certain formats. It is only since my mid-forties that the positive aspect of these experiences has begun to outweigh the negative.

"My hope is that people will keep their minds open about this subject. There's a lot more to this world than they tell you there is. I have known since I was twelve years old that there was stuff like flying saucers that they don't tell you are real. It has changed my life. My consciousness was totally altered. And UFOs are only a part of it. There is ESP and other paranormal activity. I don't know if the saucer boys magnify all of that or if they pick you because you're susceptible. But I know I am not alone anymore."

For Helen, it is critical that information become available in a free manner. She has appeared on television with Budd Hopkins to talk about the abduction phenomenon because she believes people need to be educated. "I may be a victim," she said, "but I don't have a victim mentality. It's time to pay attention to this phenomenon as a real thing. Let's look at it with open eyes. It's not going away. I think it's the most important story on Earth."

Helen Wheels is now putting together a CD anthology of old and new material that will be released in 1998 on Cellsum Records. The anthology may include some of the UFO-related songs. She will continue to write music and perform. As she expresses herself musically, she will also be helping to shift public awareness and individual consciousness about our relationship with UFOs, intelligent extraterrestrial life, and the Universe.

How does the Mindshift Hypothesis relate to popular music? In this instance, it seems that popular music about UFOs and ETs is a reflection of a change that is taking place in society, a change that may be due, in part, to the postulated covert re-education campaign. Of course, the same connections that exist between the film and television industries and government exist in the music industry as well. The corporate communications giants are inextricably linked with government on many levels.

Could rock 'n' roll music be used to convey a desired message about UFOs and ETs? From the birth of rock 'n' roll onward, conservative voices have always warned of the influence of rock music on adolescent sexuality. That message was certainly getting through. The great science fiction writer, Philip K. Dick, portrayed the subliminal manipulation of the mass mind through pop music in his novel *Radio Free Albemuth*. The ties between Washington and the entertainment industry are strong, so the posibility, if not probability, certainly exists.

However, the Mindshift Hypothesis does not require an omnipotent, all-powerful, all-controlling group that manipulates society perfectly. The hypothesis does suggest that there may be a clandestine group, comprised of imperfect human beings, with a variety of motivations, who are working behind the scenes to get UFO-related information and/or disinformation into public consciousness. Certainly, in the 1960s at least, rock 'n' roll music was a powerful force in shaping the new consciousness that was later dubbed the "youth culture" or the "Woodstock Generation."

As a young man, President Clinton may have smoked marijuana reluctantly and not inhaled, but many of the most influential rock musicians of that era enthusiastically ingested LSD and other mind-expanding or consciousness-altering drugs that greatly influenced their creative output. It is in this area that there is a connection between rock 'n' roll culture and a covert government campaign.

CIA psychochemical brainwashing and mind-control experimentation in the 1950s and 1960s included the use of LSD. There is evidence, contained in such books as *Acid Dreams: The CIA, LSD and the Sixties Rebellion* by Martin A. Lee and Bruce Shlain, that the government experimented with LSD on unsuspecting citizens in major American cities, such as San Francisco. There is speculation that some individuals within the U.S. government were even involved in the 1960s in wider distribution of LSD to the public in an attempt to see if "acid" could be used to paralyze a society and, therefore, prove effective in psychological warfare against an enemy.

In a highly ironical turnabout, it may be that covert government experiments with LSD had the opposite effect of what was intended. LSD influenced many of the most prominent musicians of the era, who in turn, produced music that opened millions of minds to the exploration of new realities, such as extraterrestrial life and UFOs. In the rock musical *Hair*, the lyrics to one song address the connection between the youth culture, the government, and LSD. The lyrics tell how President Lyndon Johnson took the IRT subway in New York City to Greenwich Village, only to see "the youth of America on LSD." The chorus intones a series of initials almost as a mantra—LBJ, CIA, LSD, LBJ—with the obvious intent of connecting them in more than a passing way.

Although LSD did not turn out to be an effective mind-control tool on a mass scale, and although the indiscriminate use of this powerful, psychotomimetic hallucinogen caused great harm to untold numbers of people, it did help shape the consciousness of many who influenced a mass audience through their music.

In addition to possible covert human activity, ET activity itself has played a significant role. The effect of the increased number of sightings of UFOs by people all over the world, and by famous musicians such as John Lennon, eventually manifested in popular music. And UFO-related songs continue to be written, performed, and heard. There is one such song, "Maybe Angels" on Sheryl Crowe's recent album. In this song, she sings, "I swear they're out there. . . . I believe they're coming back for me. . . . I'm headed down to Roswell. . . . My bag's all packed in case they come for me. . . ." Popular music continues to shape public consciousness about the UFO phenomenon.

*Michael Lindemann's* CNI News.

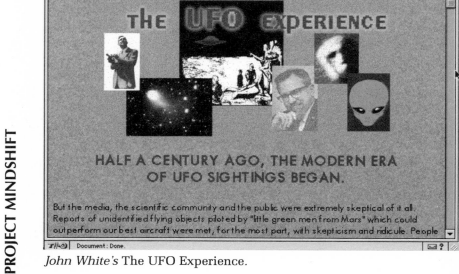

*John White's* The UFO Experience.

# CHAPTER TEN

# Mindshift on the Internet
Mindshift on the Internet
Mindshift on the Internet
Mindshift on the Internet
Mindshift on the Internet

T he way people receive the news of the world is sometimes as much of a story as the news they receive. To a generation of Americans, December 7, 1941, will always be remembered as a "Day of Infamy." And many will recall December 8, 1941, as well. Families huddled around the radio and heard President Franklin Delano Roosevelt tell the Congress and the nation that America had reached its "rendezvous with destiny." Sixty million Americans listened as the president asked Congress to declare war on Japan after the attack on Pearl Harbor. Radio was at its peak.

On November 22, 1963, President John F. Kennedy was assassinated in Dallas, the victim of a conspiracy involving at least two gunmen, according to the Senate Select Committee on Assassinations, which reached this conclusion after hearing evidence the Warren Commission did not consider over a decade earlier. All regular network television programming was suspended, and more than half of the 51.6 million homes with television sets were watching. For four days, Americans watched events unfold on television. The United States became an electronic village. Television had entered a new era.

On July 4, 1997, a new communications medium came of age. A record forty-five million people connected to the NASA Jet

Propulsion Laboratory site, and mirror sites, on the World Wide Web for news about the Mars Pathfinder mission. The Internet had reached a defining moment. During that entire week, there were *eighty million "hits" a day* on the NASA site and its twenty-one mirror sites. (A "hit" refers to the number of images and features visitors to the site reach.) A look at the Mars Pathfinder Mission site later that month revealed that the number of hits per day was still absolutely staggering.

The Internet is a gigantic network of networked computers. Anyone with a computer, a modem, and a phone can connect to the Internet. What is now called the Internet began in 1969 and was created by the Defense Department's Advanced Research Projects Agency (ARPA). The military wanted a communications system that could still function even if part of it were destroyed in nuclear war. The successful result was called ARPAnet.

In 1983, this network of networks was split into separate civilian and military systems. Other "nets" were also formed at the time, but eventually all became part of the Internet. In 1986, the National Science Foundation (NSF) created NSFNET to connect with the new high-speed computers that had been developed. NSFNET is now a core element of the Internet. Over the past few years, the NSF has been helping to create a new commercial Internet industry.

In 1985, there were about 2,000 computers linked through the Internet. In 1998, there are probably close to ten million computers linked to the Internet, supporting untold millions of additional computers. The overwhelming majority of Internet users are in the United States, but it is available to anyone in the world who has the right equipment.

A global community—or a global village, to use media prophet Marshall McLuhan's term—has been created. Through the Internet, people can communicate with one another, free from outside governmental, corporate, or media control or censorship, in an unprecedented manner. And those who value their independence on the Internet defend it fiercely.

From the beginning of the UFO phenomenon, the U.S. military has played a key role in both collecting and concealing information on intelligent extraterrestrials and their spacecraft. It was also the military that developed the communications system we know as the Internet. It is highly ironic, therefore, that the

Internet is now playing such an important role in the mindshift that is occurring in human thinking about UFOs, extraterrestrial life, and humanity's place in the Universe. The Internet is allowing UFO information about sightings, abductions, or other aspects of the subject to pass freely and almost instantaneously among all those concerned with the phenomenon.

Both serious UFO researchers and debunkers can be found on the Internet. There are also a legion of UFO sites of little or no value, filled with conscious lies, unintentional errors, and just plain old crackpot ranting and raving. In relation to the Mindshift Hypothesis, fortunately, there is as yet no way for any clandestine groups to control the flow of information in cyberspace. However, there are undoubtedly UFO websites designed by governmental, military, and intelligence agents to create confusion, doubt, and dissension through the spread of misinformation and disinformation. Of course, it would be almost impossible to discover the true identity and motivations of the individuals and organizations who post this misleading UFO data.

All Internet users are faced with the problem of finding reliable sources of information. This is especially true with controversial subjects. However, one trustworthy gatekeeper to the world of ufology on the Internet is Michael Lindemann. His website, which can be reached at http://www.cninews.com, is an excellent starting point.

Lindemann's electronic newsletter, *CNI News*, is published twice-monthly and provides an excellent sampling of UFO-related stories from around the world. Two free issues of the newsletter are offered to new visitors to his site. Some say the subject of UFOs is the second most popular site on the Internet, surpassed only by sex. *CNI News* links up with many of the better sites. Their links are listed in alphabetical order and are divided into three groups:

1. Sites devoted primarily to UFO and alien phenomena;

2. Selected sites on related topics;

3. Selected publishers and distributors of UFO books, tapes, etc.

The CNI site features the latest news about contact with non-human intelligences and also has a terrific news archive of over 900 stories (and still growing). The articles can be found by entering a query in plain English and clicking on the search command. In addition, Lindemann has compiled a "UFO Gallery" of the best documented UFO cases. There is information on the basic issues of each case, the basic ufology questions involved, and many excellent photographs. Finally, special features are added as the website is updated and improved.

One of the most useful offerings on the *CNI News* site is its "Links to UFO Websites." For those new to the subject, to the Internet, or both, this is a godsend. These links are extremely helpful to anyone trying to get his or her feet wet in the search for valuable UFO information on the Internet.

One excellent link that can be accessed through *CNI News* is sightings.com, the home page of "Sightings on the Radio" with Jeff Rense. The Sightings page contains a wide range of UFO stories, about both the most up-to-date events and classic cases. The following headlines are illustrative of the present-day stories that can be found on Sightings:

**Giant ET Craft Emerges from Sea Next to Oil Rig (Texas)**
**ET Artifacts on Moon**
**Brazil Pilot Circles Giant Mothership Three Times**
**Phoenix Councilwoman Inquires About Recent UFOs**
**Numerous UFOs Reported Over Middle East**
**Brazil UFO Forum Highlights Global UFO Situation**
**UFO Sightings Abound Across U.S.A.**

The Sightings story on the Brazilian pilot's UFO encounter contained photographs that were taken of the Mother ship. Neither the story nor the pictures received any coverage in the American press. The piece on UFO sightings across the country included descriptions of reports that took place in Indiana, Ohio, Texas, and North Carolina. UFO reports from Central America, Mexico, and Canada also were covered recently. In combination with *CNI News*, the Sightings page offers a timely, in-depth look at the UFO phenomenon around the world.

There are also classic cases on the Sightings homepage. For example, there is a report on the famous case of Thomas

Mantell, Jr., the fighter pilot who died on January 7, 1948, while chasing a large UFO. Although the official version claims that this experienced flyer was chasing Venus, then barely visible as a tiny point of light below Mantell's plane, his radio communications indicate he was closing in on a very large craft flying above him. Strangely, *the official reports and radio transcripts of the Mantell case remain classified fifty years after the event.* Why? If there is proof Mantell was mistakenly and fatally chasing Venus, why not release the official reports and transcripts?

Other fascinating items on the Sightings homepage include pieces on ET intervention in human evolution; Laurance Rockefeller's UFO activities; Disney's sneak preview of a UFO documentary in five U.S. cities; and a U.S. Navy recommended UFO reading list. There is also a link from the "Sightings on Radio" page to the "Sightings on TV" page.

Lindemann's UFO homepage also contains links to other intriguing websites devoted to UFOs and alien phenomena. For example, visitors to cninews.com can click on Budd Hopkins's "Intruders Foundation" homepage. This site includes details of upcoming appearances by Budd Hopkins; late-breaking UFO news stories; and a connection to the Abduction Information Center, which features a Bulletin Board, chat room, and an interactive site for online research reports.

*CNI News* also provides a direct link to such major sources of information as the J. Allen Hynek Center for UFO Studies (CUFOS); Mutual UFO Network (MUFON); Fund for UFO Research (FUFOR); MUFON Canada; and many other equally valuable and interesting sites. The page is well-organized and simple to work with. It is worth a visit by anyone seeking UFO information online.

Art Bell's homepage is also worth visiting. This commercial site is devoted to Art Bell's radio talk show and includes many more topics than UFOs and extraterrestrial life. His show is one of the best around when it comes to investigating the UFO phenomenon. He treats the subject seriously and is fair with the guests. In addition, the show frequently features a number of ufologists who engage in vigorous debate live on the radio. This site features an art gallery with space images, photographs of UFOs, and images of alien entities.

James Hurtak's "Academy for Future Science" homepage is one of the most scientific pages on the subject. Hurtak's scientific research is presented in a straightforward fashion. In addition, the site offers information on what the Academy is; provides access to the journal *Future History*, devoted to cross-disciplinary exchange among scientists, educators, and other professionals; gives "Science Flash" briefings on current hot topics, such as Mars explorations; and contains abstracts of current research, complete with references.

FIONS, or The Friends of the Institute of Noetic Sciences, http://www.fions.org, has a homepage that is worth visiting as well. This small, New York City-based organization has sponsored some remarkable events involving experts in UFO-related matters. On a modest budget, FIONS sponsors and produces events of such high caliber that they rival the efforts of much larger organizations.

For example, Dr. John Mack spoke about his work with abductees, or "experiencers," at a FIONS event in April 1997; Apollo 14 astronaut Dr. Edgar Mitchell gave a day-long presentation on his work on November 22, 1997; and James Hurtak delivered a talk in March 1998 on the scientific revolution that is underway, in which he addressed such topics as extraterrestrial life and zero-point energy. Dr. Hurtak gave a brilliant presentation on extraterrestrial life and UFOs. In his talk, Dr. Hurtak showed astounding video images of UFOs from Russian television news and from a home video showing a UFO over New York harbor. His talk was extraordinary. Dr. Hurtak expressed views that are in accord with the Mindshift Hypothesis during his talk. In addition, on October 31 and November 1, 1997, FIONS held a centennial symposium celebrating the life and work of Wilhelm Reich at the prestigious New York Academy of Medicine. This event shattered a silence that has surrounded Reich's science since his tragic death in a Federal penitentiary in 1957. Among the speakers that weekend were Mary Higgins, the director of the Wilhelm Reich Museum; Dr. John Mack; *Left at East Gate* author Peter Robbins; and the author of this book as well.

The UFO Experience (http://www.ufoexperience.com) is an excellent Internet resource for information about UFOs and extraterrestrial life. It is the homepage of John White, an author on such topics as UFOs and extraterrestrial life, consciousness research, and higher human development. White has written fifteen books, among them *The Meeting of Science and Spirit*. This homepage is dedicated to providing "information for transformation"; that is, knowledge that advances humanity's understanding of human and cosmic history, the possibility of other intelligent life in the Universe, and our relationship to the Universe in its many aspects.

The site features information about White's superb annual UFO conference, at which eminent investigators speak. The conference offers a truly comprehensive look at the UFO phenomenon. Detailed information on speakers, registration and other related matters are available. One can also access information concerning audio and video tapes of past conferences, the UFO Experience book series, and the UFO Experience television series. In addition, the site provides links to other fascinating UFO-related websites.

On June 30, 1997, the *New York Times* published an article, "UFO Believers and Debunkers Thrive on the Web." True to form, this debunking article covered the topic in a derisive, mocking tone. The piece praised the usual suspects: the Air Force, debunker Philip Klass, and the Committee for Scientific Investigation of Claims of the Paranormal (CSICOP). Predictably, the article leaves the clear impression that all who take the UFO phenomenon seriously are "crackpots," "conspiracy theorists," "true believers," "pseudoscientists," "paranoids," "wild-eyed ufologists," "hucksters," or "charlatans." According to the *Times*, UFO debunkers engage in "carefully reasoned analysis" and provide badly needed "reality-check sites" online, while those who take the subject seriously "rant" or "spread rumors, fantasy and just plain fiction," or "purvey dubious space alien stories." It is amazing to read a short article so chockfull of perjoratives concerning UFO investigators or individuals who have an open mind on the subject.

The *Times* article provides the reader with the Internet addresses of eighteen UFO-related websites, thirteen of which are maintained by debunkers of one sort or another. Although the powerful in government, intelligence, and mass communications cannot control the Internet, they are a presence on it. And through their own websites the major newspapers and magazines, the television networks and large cable organizations, and other corporate and political interests get their messages across. Fortunately, they cannot yet dominate cyberspace, try as they might.

The Internet is like swimming; you have to dive right in. It may take time at first to find what you are looking for, but that improves with experience. Also, the technology itself can sometimes be excruciatingly time-consuming. (The World Wide Web is often jokingly referred to as the World Wide Wait.) However, it is fascinating to explore the wealth of information available on UFOs and related subjects, such as space exploration. It is also possible to communicate directly with people on every continent who share similar interests. And information that is censored, or under-reported, by the mainstream media finds an outlet on the Internet.

There is a low, steady drumbeat of criticism of the Internet in many quarters of the mass media today. Newspaper and magazine reporters and editors, columnists and correspondents, television anchormen and Sunday morning talking heads, all complain about the lack of journalistic professionalism on the Internet, the lack of controls on expression. To those who value the Internet, the inability of the powerful media conglomerates to shape and censor this new medium is a great boon to true communication.

There are excesses, errors, and irresponsibility on the Internet, to be sure. But the media stars attacking the Internet are like people living in glass houses. The so-called news has degenerated into entertainment—celebrity puff pieces, promotional pieces, or "program notes" for entertainment shows and exploitative, tabloid coverage of inconsequential but lurid events. Money, not journalism, determines the mass media's focus. For example, as this is being written, the highly paid network news readers are tripping over one another to get out of Cuba, abandoning coverage of the Pope's visit to Castro's Cuba, which they hyped for weeks as historic, in a humiliating bid to grab higher ratings by covering a sex scandal instead. Newspapers, magazines, and radio are also saturated with this possible White House sex scandal. At the very same time, an impending war with Iraq and a spreading financial crisis throughout Asia (which has the potential to lead to a worldwide economic depression) are being ignored. This is the journalism that is extolled as being superior to the best of the Internet.

The following list is composed of websites that provide a wide range of information and opinions about UFOs, extraterrestrial life, space exploration, consciousness studies, shifting paradigms in science, and other related topics. The information is not censored. The information is not slanted because of corporate policy. However, each person who surfs the net will have to use his or her own judgment in determining what information is valuable and which sources are reliable. Many of the sites below may be accessed through cninews.com.

## UFO and Alien Phenomena:

Abductees Anonymous
Anomalous Images and UFO Files
Budd Hopkins' Intruders Foundation
BUFORA—British UFO Research Association
CISU (Italy) UFO INFO
CSETI—Center for the Study of Extraterrestrial Intelligence
CUFOS—J. Allen Hynek Center for UFO Studies
Fire in the Sky—Travis Walton
Flying Saucer Review
FIONS—Friends of the Institute of Noetic Sciences
FUFOR—Fund for UFO Research
MUFON—Mutual UFO Network
MUFON Canada
National UFO Reporting Center
Rutgers University UFO Archive
The UFO Experience
UFO Sweden
WWW Virtual Library: UFOs
Yahoo on UFOs

## Sites with UFO or Alien Topics:

Academy for Future Science
Art Bell's Home Page
The Farsight Institute (Dr. Courtney Brown)
NASA Homepage
Sightings on the Radio—with Jeff Rense
The SETI Institute

## Sites Recommended by the *New York Times*:

UFORIA
International UFO Museum and Research Center, Roswell
The Roswell Report Case Closed (1997)
The Roswell Report (1994)
Bruce D. Kettler's Page
60 Greatest Conspiracies
Parascope
Circlemakers

The Internet is a world of its own. As the poet e. e. cummings wrote:

> there's a hell of a universe next door
> —let's go!

*The development of Wilhelm Reich's science may bring about the greatest mindshift of all.*

# CHAPTER ELEVEN

# Mindshift in Science
# Mindshift in Science
# Mindshift in Science
# Mindshift in Science
# Mindshift in Science

S cience in the twentieth century has undergone a transition from the view that *matter* is the fundamental property of Nature to a conception of *energy-matter* as the fundamental physical reality. This is an intermediary stage. We are now on the verge of even further progress toward an understanding of physical reality in which *lawfully functioning mass-free energy* is proven experimentally to be fundamental. The *process* of comprehending the lawful functions of mass-free physical energy, and of confirming its existence scientifically, will of necessity initiate a gradual and complete reshaping of human character structure and human society.

There are indications that the scientific investigation of mass-free energy was blocked in the 1950s, possibly by the same covert forces that were active in controlling information about the extraterrestrial phenomenon. Although it is not *yet* widely recognized, a profound mindshift in science, toward this new view of reality, is now underway. It cannot be stopped.

A new generation entering our world—the children of the future—will see this profound change come to fruition. As adults, these children will undertake the scientific experimentation that will confirm the existence of the physical, mass-free,

A Sunburst Book. Farrar, Straus and Giroux, 1989.

Chronicle Books, 1997.

Usborne Publishing, Ltd, 1996.

primordial energy that pervades the Universe. The first steps toward this reality are already being taken. It can be seen in something as ordinary as children's books. A number of books, marketed by their publishers for youngsters aged three to eight or five to nine, have been published in the last year or two that present the cosmos to children in understandable ways.

*Here in Space*, written and illustrated by David Milgrim, opens with the lines, "I live in space, I have lived here since birth,/On a big ball of rock that is called planet Earth." In the book, a young boy and his dog travel all over our planet and meet all kinds of "alien" life right here on Earth. The young boy is a *planetary citizen* who travels freely on Earth and is not limited by national boundaries. The text and drawings will both satisfy and stir the imagination of young readers about who they are and their place in the Universe. *Man on the Moon*, written by Anastasia Suen and illustrated by Benrei Huang, provides a factual account of the achievements of the Apollo 11 NASA mission to the Moon. *Floating Home*, by David Getz, tells of a girl's dream of becoming the first eight-year-old astronaut. In addition to being a wonderful story of her fantasy of life as an astronaut, the book introduces factual information about the realities of space travel.

Three other children's books involve more than human exploration of outer space. They deal with UFOs and intelligent extraterrestrial life. Published in 1989, *UFO Diary*, by Satoshi Kitamura, was ahead of his time. The simple text, beautifully illustrated, is narrated by an extraterrestrial explorer who opens his tale, "On Monday, I took a wrong turn in the Milky Way." The kindly, curious ET comes to Earth, where he befriends a young boy who shows him the beauty of this planet. The alien returns the favor by taking the child in his ship far enough out into space so the boy can see the entire Earth. In this beautiful book, each being has love and respect for the other. *UFOs*, edited by Caroline Young, was published in the United States in 1996. It is an excellent introduction to ufology and human space exploration for children—and even adults. In the 1997 book *Vacationers from Outer Space* by Edward Valfre, a family car trip becomes an intergalactic adventure. The book begins, "Once we took a family car trip into the middle of nowhere. . . . In a flash of light, a flying saucer appeared above our table. Who were they? Where did they come from?" Valfre's text and photography are witty and surreal, but the UFO seems natural enough.

In books such as these, children are not being introduced to outer space as a battlefield, as another arena in which human gladiators can engage in violent contests. Youngsters are being shown a Universe that is open to exploration and that is most likely filled with other searching, intelligent lifeforms. The immature days of Buck Rogers, Flash Gordon, and Star Wars are gradually yielding to a mature view of humanity's emergence into the cosmos—not as a conqueror or combatant, but as an organism reaching a new level of consciousness. Many adults will resist the new scientific truths and the change they bring about in our worldview. To a new generation of children, the emerging worldview will be the only reality they know.

As Max Planck wrote in his *Scientific Autobiography*, "A new scientific truth does not triumph by convincing its opponents and making them see the light, but rather because its opponents eventually die, and a new generation grows up that is familiar with it." We are truly living through a paradigm shift.

The term "paradigm shift," which is currently so popular, was introduced by Thomas Kuhn in his 1962 book, *The Structure of Scientific Revolutions*. Kuhn's book is an amplification of a project that Kuhn, later the Laurance S. Rockefeller professor of philosophy at the Massachusetts Institute of Technology, had conceived fifteen years earlier as a graduate student in theoretical physics. The word *paradigm*, as with all terms that become popularized and used as "buzzwords," has lost much of its meaning through misapplication. We hear of paradigm shifts through such empty enterprises as advertising, marketing, and public relations campaigns in a manner that deprives the term of its significance. What exactly is a paradigm?

Through his study of the history of science, Kuhn came to recognize what he called a paradigm: "the entire constellation of beliefs, values, techniques and so on shared by members of a given community." In relation to science in particular, Kuhn defined a paradigm as "universally recognized scientific achievements that for a time provide model problems and solutions to a community of practitioners."

In his book, *Unconventional Flying Objects: A Scientific Analysis*, Paul R. Hill, a NASA scientist, saw the clear connection between Kuhn's observations and the manner in which the scientific community has so far treated the UFO phenomenon. To Hill, anyone who has read Kuhn's book "must know that the acceptance of the UFO has to be the gradual process that it is turning out to be because it is man's nature, and scientific history, that old ideas are discarded only after new ideas are firmly established. It often takes new generations to squarely face new facts." Humanity is now living through a transitional phase, the outcome of which is uncertain.

However, the effects of the mindshift can be seen in the press coverage of the possibility of life elsewhere in the solar system and the Universe. The current spate of headlines stands in stark contrast to stories about the UFO phenomenon. Why is the possibility of life on other planets and moons treated in a rational manner, whereas the coverage of the possibility of intelligent extraterrestrial life is so slanted? Editorial policies are consciously set and maintained. Is there covert collaboration between some in the media and a clandestine group in government concerning these related subjects? A random sampling of headlines from the *New York Times* in the two years make it clear that the two topics are treated quite differently:

Scientists Discover New Planet, Creating Hopes of Finding More
In a Golden Age of Discovery, Faraway Worlds Beckon
Jupiter's Moon Might Be Cradle for New Life
Data on Jupiter Moons Offer Tantalizing Hints of Possible Life
Scientists Widen the Hunt for Alien Life
Clues in Meteorite Seem to Show Signs of Life on Mars Long Ago
Skeptics Are Impressed as Mars Rock Is Shown
Study Backs Idea That Meteorite Hints of Life on Mars
New Traces of Past Life on Mars
Three Mars Layers Suggest Life on Planet Was Possible
NASA Still Dreams of Mars Outpost

On February 9, 1997, John Noble Wilford, a science writer for the *New York Times*, observed in a page-one story, that "All of a sudden, astronomers have turned a big corner. . . . Other worlds are no longer the stuff of dreams and philosophic musings. They are out there, beckoning, with the potential to change forever humanity's perspective on its place in the universe." The article also points out that, in the next decade, space telescopes will be available that "should be able to see planets as small as Earth elsewhere and examine their atmospheres for signs of life." There are one hundred billion stars in the Milky Way alone and *half of them may have planets*, many capable of supporting life, including intelligent life.

Today, we know of more planets outside of our solar system than within it. How close are human beings to exploring these planets? A spacecraft called *Kepler* may be launched in March 2001 to study these new worlds. Within a few weeks, it could detect thousands of new planets, many of which would be similar to Earth and on which life could have arisen. The space technology aboard the craft is so sensitive that, from orbit, it could detect one person passing a flashlight to another on the surface of the planet. Daniel S. Goldin of NASA has said, correctly, that the discovery of life on other planets will leave no aspect of human existence unchanged. Scientists now suspect that as many as ten bodies in our own solar system harbor some form of life.

The old mindset is indeed shifting. Recent findings about Mars are causing a revolution in the field of exobiology, the study of life in the cosmos beyond Earth. NASA is now hiring biologists to help develop programs to search for life elsewhere. This is a radical change from 1976, when the last NASA probe landed on Mars. At that time, *chemists*, not *biologists*, won out in the interpretation of the experiments conducted on the Martian soil. The chemists concluded there were no indisputable signs of life.

NASA is also encouraged by the evidence of signs of primitive life found on a meteorite believed to have come from Mars. In addition, recent findings suggest that microbial life may have begun at the same time—3.6 billion years ago—on both Earth and Mars. According to an Associated Press report on October 8, 1997, the Mars Pathfinder data provided strong evidence that Mars was once warm and wet enough to be hospitable to life.

In the past, progress was slow. Change is now occurring at an extraordinary pace. In 1609, Kepler proposed his three harmonic laws of planetary motion. It took 360 years for mankind to progress to the point where human beings could use those laws to send spacecraft to other worlds. In 1903, Orville and Wilbur Wright flew the first successful plane at Kitty Hawk, North Carolina. It took only sixty-six years to move from a rudimentary airplane to the 1969 Moon landing. In 1976, NASA scientists concluded that there were no signs of life on Mars. In 1996, signs of Martian life were everywhere. Only a few years ago, scientists had not located another planet beyond our solar system. Soon, the discovery of new planets will be an ordinary event.

The mindshift is not only obvious in newspapers. Just as children's minds are being enlivened and opened by books about the cosmos and the UFO phenomenon, adult minds are being challenged and changed as well. Many excellent books on the new cosmologies have been published recently and they are advancing this positive mindshift.

## Mindshift Reading

*Cosmology and Controversy*, by Helge Kragh, professor of the history of science at the University of Oslo, is a thorough, serious study of the development of scientific cosmology and the people who created our new view of the Universe. Kragh details how, from 1920 to 1970, cosmology became a science—a branch of physics—and was no longer confined only to myth, religion, and philosophy. The author focuses on the two main cosmologies of our century, the *Steady State Theory* and the *Big Bang Theory*.

The Steady State Theory is a model of dynamic equilibrium, in which there is continuous creation through a perpetual exchange of energy between matter and radiation in a stationary Universe of infinite age. In contrast, the Big Bang Theory asserts that a superdense state of matter exploded and created an evolving Universe that had a beginning in time and is of a finite age.

Alan H. Guth is the author of *The Inflationary Universe* and the V.F. Weisskopf professor of physics at MIT. *Newsweek* has called him "one of the top 25 American innovators," and *Science Digest* has listed him as "one of the 100 brightest scientists under 40." In his book, he asks three provocative questions, "What happened *before* the Big Bang? What actually banged? What caused it in the first place?"

To Guth, the Big Bang is actually an explanation of the *aftermath* of the bang, not of the bang itself. To explain what happened in the *billionth-trillionth-trillionth* of a second that was the Big Bang itself, Guth proposed the Inflationary Universe Theory, which is now widely accepted. Stephen Hawking was an early supporter of this theory, and data from the U.S. space probe COBE seem to confirm Guth's hypothesis. In Guth's theory, the Universe began with a rapid expansion in an area smaller than an atom that contained only one ounce of mass. There was a rapid expansion, or inflation, followed by the bang, and then less rapid expansion ever since. From this point on, for Guth and others, the theory of the Big Bang explains the origin of the Universe. Although the Big Bang theory holds sway today, it has its critics.

Eric Lerner, author of *The Big Bang Never Happened*, asserts that the explosion that occurred ten to twenty billion years ago—and which is central to modern cosmology and all current physical theories of energy and matter—is a modern myth. According to Lerner, the Big Bang Theory shares much in common with Medieval cosmology because both posit a finite cosmos created *ex nihilo*, out of nothing. He believes that the "Plasma Cosmology" of Hannes Alfven and Ilya Progogine—which describes an evolving universe, infinite in both time and space—is a more likely explanation of our physical reality.

In *The Life of the Cosmos*, by Lee Smolin, professor of physics at the Center for Gravitational Physics and Geometry at Pennsylvania State University, Smolin describes the sweeping, profound revolution that is occurring in cosmology and asks three eternal questions: "What is the Universe? What is Life? What is a human being?" He struggles to unify quantum theory, relativity theory and cosmology, and presents the reader with a new understanding of the Universe as an interrelated, self-constructed entity. Smolin's physics is an attempt to unify Einstein's theory of general relativity and quantum mechanics to create a new cosmology.

Each of these books is an example of the paradigm shift that is taking place in cosmology and physics. We are also seeing the beginnings of a change in the attitudes of some in the scientific community regarding the UFO phenomenon. According to Dr. Michael E. Zimmerman, a professor of philosophy and former chair of the Department of Philosophy at Tulane University, some mainstream scientists are finally beginning to take the UFO and abduction phenomena seriously and to study them systematically. In a talk he gave at the Omega Institute, Dr. Zimmerman noted that a report by the Brookings Institute in the 1950s, "The Implications of a Discovery of Extraterrestrial Life," suggested that "of all groups, scientists and engineers may be the most devastated by the discovery of relatively superior creatures." However, in a recent survey, 75 percent of scientists said they would like to learn more about the UFO phenomenon. Zimmerman believes that many scientists refuse to discuss UFOs in public for three main reasons:

1. fear of loss of social staus through ridicule;
2. superior non-human intelligence threatens personal psychology and worldview; and/or
3. fear of social chaos in the face of superior ET reality.

In his paper, Zimmerman also discussed the suspicions of some ufologists that secret government agencies leak UFO information for their own purposes and may even encourage the production of mass entertainment *designed to prepare the public* for the revelation that intelligent ETs are visiting our planet. Zimmerman does not support what this book calls the Mindshift Hypothesis, but does acknowledge that there could be a covert effort by the government to re-educate the public about UFOs or release information in ways that serve its own agenda.

A number of books about UFOs and science should be of interest to scientists and nonscientists alike. Some are classics, others are new, but they help bridge the gap between science and ufology. *UFOs: A Scientific Debate* is a collection of essays edited by Carl Sagan and Thornton Page. It offers the reader a good, representative mix of scientific opinion on the subject. *At the Threshold: UFOs, Science and the New Age* by Charles F. Emmons, is an interesting contribution to the debate, as is *Science and the UFOs* by Jenny Randles and Peter Warrington. *Is Anyone Out There?* by Frank Drake of SETI, and Dava Sobel, is a highly readable, thought-provoking book. To Drake, it is only a matter of when we will meet the ETs, not *if.* J. Allen Hynek's contribution, *The UFO Experience: A Scientific Inquiry*, is a classic, and his thoughts are deserving of serious consideration.

Paul R. Hill's unique, exciting study, *Unconventional Flying Objects: A Scientific Analysis*, is the valuable contribution of a NASA scientist who for years was an unofficial clearinghouse for UFO information within NASA. This book contains excellent information on the underlying science and technology of UFOs. Hill's early reports of humanoid occupants and *communication with them* predate the abduction phenomenon. Dr. Edgar Mitchell has written of this book, "Paul Hill has done a masterful job ferreting out the basic science and technology behind the elusive UFO characteristics *and demonstrating that they are just advanced and exotic extensions of our own technologies.* Perhaps this book will help bring solid consideration for making all that is known about extraterrestrial craft publicly available."

Although the scientific community is only slowly beginning to look seriously at the subject of UFOs, there are other fields of scientific investigation that are bringing the scientific community closer to areas of inquiry that interest the UFO community. One of these topics is the study of *zero-point energy.*

## Zero-Point Energy

What is zero-point energy? The American Heritage Dictionary describes zero-point energy as "the irreducible minimum energy possessed by a substance at absolute zero temperature." NASA defines it as "kinetic energy retained by molecules of a substance at a temperature of absolute zero." These basic definitions describe aspects of the energy, but do not indicate why it is significant.

According to Dr. Edgar Mitchell, science now has shown that the basic structure of matter contains two things—empty space and energy. And even the empty space is now assumed to contain energy, the so-called "vacuum energy" or the zero-point field. In his book, *The Way of the Explorer*, Dr. Mitchell writes that "The zero-point field is defined as the field of quantum fluctuations that exists at a temperature of absolute zero. *More recently, however, it has also been interpreted as that field of energy that underlies all matter.*" The italics have been added because of the significance of that statement. To Dr. Mitchell, the zero-point field is the basic, unstructured, quantum potential from which the Big Bang arose. In fact, everything arose from the zero-point field of energy.

Zero-point energy is ubiquitous and fills the fabric of all space. The existence of zero-point energy was discovered in 1958 by a Dutch physicist, M.J. Sparnay, who continued the 1948 experiments of Hendrik B. G. Casimir. These experiments showed that there were physical forces between two uncharged parallel metal plates in a vacuum. Sparnay discovered that the forces were the result of a type of radiation now called zero-point energy, which exists in a vacuum and is present even at a temperature of absolute zero. In January 1997, in a paper published in *Physical Review Letters* (Vol. 78, No.1, pp. 5–8), Steven K. Lamoreaux reported on his experiment confirming the Casimir effect and the existence of zero-point energy.

A vacuum at absolute zero is no longer considered to be empty of all electromagnetic fields in physics. Rather, the vacuum is seen as containing random, intense, fluctuating zero-point energy fields in which energy and matter are constantly changing from one state to the other.

Soviet physicist Andrei Sakharov thought that gravity itself may be a result of fluctuations in the zero-point energy field. Sakharov conceived of all matter as floating in an energy sea. Today, physicists know that all space—that between the stars and that between the particles that make up matter—is filled with fluctuating zero-point energy. Zero-point energy is inherent to the fabric of space. Quantum mechanics effects may arise from the interaction of zero-point energy and matter. Throughout the twentieth century, physicists have sought in vain for a unified field theory to explain the physical world. The zero-point energy field may point the way to a solution. But only experimental science can provide the proof that is needed.

If we are all floating in a sea of energy, with a vast, untapped potential, can this energy be used practically? Not all scientists agree. For example, John Obienin, a researcher at the University of Nebraska at Omaha, thinks that the idea is too good to be true. Although he recognizes that what is called a vacuum is actually *seething with energy even when no matter is present,* Obienin asserts that, at present, no one knows how to use zero-point energy to deliver a substantial amount of power. John Baez, a mathematician at the University of California at Riverside, does not take zero-point energy, or "vacuum energy," literally. To him, it is simply a mathematical tool; it is not a physical reality. Paul A. Deck, an assistant professor of chemistry at Virginia Polytechnic Institute and State University, claims that zero-point energy cannot be harnessed as a power source at all.

However, various experiments conducted by other scientists do indicate that zero-point energy has practical applications, such as described in Robert Forward's 1984 paper, "Extracting electrical energy from the vacuum by cohesion of charged foliated conductors." In 1993, Cole and Puthoff presented their work with zero-point energy in a paper, "Extracting energy and heat from the vacuum." A U.S. patent was issued to F. B. Mead, Jr., and J. Nachamkin on December 31, 1996, for a "System for converting electromagnetic radiation energy to electrical energy." In 1993, echoing Sakharov, Puthoff published a paper, "Gravity as a zero-point fluctuation point."

Nikola Tesla wrote, in 1891, "Throughout space there is energy. Is this energy static or kinetic? If static, our hopes are in vain. If kinetic—and this we know it is, for certain—then it is a mere question of time when men will succeed in attaching their machinery to the very wheelwork of nature." It seems highly probable that science will be able to tap the virtually limitless quantities of zero-point energy directly. However, much hard work remains to be done concerning two theories: one, the theory that the *vacuum contains real energy fluctuations*; and, two, theories of system self-organization. But the possibilities are astounding.

If turned into reality, a zero-point energy technology would signal the end of the fossil-fuel era, a fact that cannot have escaped the attention of those for whom oil is their lifeblood.

Zero-point technology could even be used in spaceships. These ships, powered by the energy of the Universe itself, would not need propellant. NASA scientists and others around the world are now exploring this possibility. For example, H. E. Puthoff, Ph.D., director of the Institute for Advanced Studies in Austin, Texas, wrote a paper in September 1997 called, "Can the Vacuum Be Engineered for Spaceflight Applications?"

To Puthoff, empty space is not empty at all. It is "the seat of myriad energy quantum processes that have profound implications for future space travel." This has not gone unnoticed by the U.S. Air Force, which has solicited research proposals for its Air Force Rocket Propulsion Laboratory to fund in the areas of *"esoteric energy sources for propulsion including the zero-point quantum dynamic energy of vacuum space."* Robert Forward submitted a study to the Air Force in which he stated that "Many researchers see the vacuum as a central ingredient of twenty-first century physics. Some even believe the vacuum may be harnessed to provide a limitless supply of energy."

Can the vacuum be engineered for spaceflight applications? The answer to Puthoff's question is, in principle, yes. But there is a long way to go before a zero-point energy technology becomes a practical source of power.

In terms of the Mindshift Hypothesis, Dr. Puthoff is as interesting a case as his theories. He graduated from Stanford University in 1967 and since then has been affiliated with General Electric, Sperry, his alma mater Stanford, SRI International, and now with the Institute he heads. Dr. Puthoff has been a regular consultant to the Executive Branch, Congress, and to many government agencies. In addition, he has been involved with the National Security Agency. Dr. Puthoff has worked in the corporate, educational, governmental, and intelligence milieu that forms the core of the so-called military-industrial complex and the National Security State. His thoughts on zero-point energy technology and its applications must be considered in the context of this background.

Human beings are now investigating the use of the energy of space itself. How does this relate to UFOs and extraterrestrial technology? Ufologists have long speculated that ET spacecraft may employ a technology that uses the energy of space. This likelihood seems greater in light of humanity's own developing space science. Changing paradigms in physics and cosmology, as well advances in space science, are opening minds to new realities and new possibilities, bringing mainstream science and ufology closer than either group may be aware of or comfortable with.

Aspects of the relatively new research into the phenomenon of zero-point energy bear a striking resemblance to the scientific energy investigations of Wilhelm Reich, the pioneering scientist who has been mentioned on a number of occasions throughout this book. Reich's radical impact on science and society is just now being recognized by a large and growing number of people all across the United States and around the world. His energy investigations led him from the study of the human psyche and human character structure to the exploration of the cosmos and consideration of the possibility of contact with extraterrestrial life.

It is important to know something of Reich's background before considering his contributions to the new physics and the new cosmology, as well as to our understanding of the UFO phenomenon.

## Wilhelm Reich: Pioneer of the Paradigm Shift

Wilhelm Reich was born on March 24, 1897, in the easternmost region of the Austro-Hungarian Empire. He grew up on a farm and lived close to Nature. His early experiences in this natural environment helped shape his contributions to psychoanalysis, medicine, and science. In the 1920s, Reich was considered by many to be the "heir apparent" to Freud. In the 1930s, his work on biogenesis—the origin of life—led some to believe him worthy of a Nobel Prize. In the 1940s, he discovered a new form of energy and developed a new "bodymind" energy medicine. In the 1950s, Reich elaborated a new paradigm in physics, one in which energy, and not matter, is primary.

On November 3, 1957, Wilhelm Reich died in a Federal penitentiary, his medical and scientific books burned by the U.S. government. Today, few people have heard of Reich; even fewer are aware that he died in prison and that his books were burned. However, despite the book burning, his work has survived and a new generation is eagerly becoming acquainted with it.

Reich introduced a radically new concept of the Life Energy into human thinking. Energy was no longer a mystical or philosophical concept. It was a tangible, measurable, usable physical energy. He called the energy "orgone" because it was discovered through his study of the orgasm and because it charged organic matter.

He investigated energy functions, from his 1923 psychoanalytic paper "Concerning the Energy of Drives" to his 1957 book *Contact with Space*, his last work on cosmic energy functions in man and nature. Reich's energy investigations are "the red thread," as he called it, that runs throughout his work. The body of work that Reich left is a treasure waiting to be claimed.

In 1920, while still a twenty-three-year-old medical student, Reich became a member of Freud's Vienna Psychoanalytic Society. In 1922, he earned his medical degree from the Medical School of the University of Vienna. Reich worked with Nobel Prize winner Professor Wagner-Jauregg at the Neurological and Psychiatric University Clinic. Few students have been as fortunate as Reich, who studied with a Nobel Laureate in neurology and with Sigmund Freud, the founder of psychoanalysis.

In the 1920s and 1930s, Reich founded free mental health clinics for poor people. He supported these clinics with his own earnings. Reich marched in the streets with workers and fought for women's reproductive rights and economic independence. He was in the forefront of the movement to protect the sexual rights of infants, children, and adolescents. In addition, Reich held mass sexual education rallies at which men and women received direct answers to the burning sexual questions that troubled them deeply. He was active in the anti-Nazi underground and was an outspoken critic of both Hitler and Stalin.

In 1933, after years of painstaking clinical work in psychoanalysis and mass psychology, Reich published two classic works—*Character Analysis* and *The Mass Psychology* of Fascism. That same year, he fled Germany and went to Scandinavia when Hitler assumed complete power. In 1934, Reich was expelled from both the International Psychoanalytic Society (IPA) and the Communist Party because his ideas were unacceptable to psychoanalytic and political functionaries. His forthright stands on life-and-death issues in science and society made him many enemies. Through his mass psychological work, Reich came to a radical insight: he saw that the problems of society lay in the character structure of the average human being.

In Scandinavia, in the 1930s, Reich explored new areas of research and developed a new body-mind therapy. At this time, he performed a series of experiments, described in his book *The Bioelectrical Investigation of Sexuality and Anxiety*, in which he measured an energy charge at the skin surface and at the erogenous zones in human beings.

During these fruitful years, Reich also performed a series of pioneering microbiological experiments in which he discovered radiating energy vesicles he called "bions" (from the Greek word for "living"). He found that the bions represented a transitional stage between living and nonliving matter. He believed that bions form constantly in nature through a process of disintegration of organic and inorganic matter and that it is possible to reproduce this process experimentally. In the late 1930s, Roger Du Teil of the French Academy of Sciences confirmed Reich's bion research experiments into the origins of life.

Reich was a pioneer in time-lapse cinematography and recorded his bion experiments on film, segments of which can be seen at The Wilhelm Reich Museum in Rangeley, Maine. This work is described in *The Bion Experiments on the Origin of Life*.

From 1930 to 1939, Reich was on the move—from Vienna to Berlin, Malmo, Copenhagen, and Oslo. In the summer of 1939, Reich had to flee for his life from the Nazis. As World War II approached, Reich was invited to teach at The New School for Social Research in New York City, where he was appointed associate professor of medical psychology.

In the summer of 1940, while on vacation in Maine, Reich made his first observations of the atmospheric orgone energy. His investigation of orgone energy led to the development of a number of medical and scientific devices, such as the orgone energy accumulator; an orgone energy field meter; an orgone energy motor; the cloudbuster, a weather modification device; and the Medical DOR-buster, a medical device that withdraws stagnant biological energy from rigid musculature.

Controversy began to swirl around Reich shortly after the English language edition of *The Mass Psychology of Fascism* appeared in 1945. From 1947 to 1957, the FDA secretly investigated and then prosecuted Reich for both his ideas and the influence of those ideas. The FDA campaign ended with a trial held in Portland, Maine, in 1956. *For a decade, the FDA spent between 4 and 6 percent of its entire budget to investigate and prosecute Reich*—inexplicable behavior that defies logical explanation. As late as February 1963, six years after Reich's death, the FDA was still concerned that his books were available in paperback and upset that "there is still considerable support for his theories."

A jury found Reich guilty of criminal contempt in disobeying a 1954 court injunction against his medical and scientific work. He was sentenced to two years in prison and fined ten thousand dollars. The FDA destroyed Reich's orgone energy accumulators in Maine on June 5 and July 23, 1956. On June 26, 1956, again in Maine, the FDA burned 251 pieces of Reich's scientific literature. At that time, Reich told a government agent that his books had been burned by the Nazis in Germany, but that he had not expected it to happen again in the United States.

In New York City, on August 23, 1956, six tons of Reich's books were burned by the FDA. Among the works consigned to the flames that day were: *The Mass Psychology of Fascism*; *The Sexual Revolution*; *People in Trouble*; *Character Analysis*; *The Function of the Orgasm*; *The Cancer Biopathy*; *Listen, Little Man!*; *The Murder of Christ*; *Ether, God and Devil*; and *Cosmic Superimposition*. In conducting these book burnings, zealous FDA agents went beyond the court order and, consequently, acted without authority. There is even evidence that FDA files regarding Reich were falsified in a cover-up of these illegal activities.

In the decade before his death, years in which he was harassed and his work was disrupted by the legal campaign of the U.S. government, Reich continued to work vigorously, pursuing the "red thread" of energy functions wherever it led. Reich found himself entering unknown territory once again, and on the verge of establishing a new paradigm in physics.

Reich's earlier work had passed through Schopenhauer's Three Stages of Truth: "First, it is ridiculed. Second, it is violently opposed. Third, it is accepted as self-evident." However, Reich's core discovery—the orgone energy—remained taboo.

In 1982, with the publication of *The Turning Point* by Fritjof Capra, Reich's orgone physics came under serious consideration by a noted physicist. Capra wrote, *"Wilhelm Reich was a pioneer of the paradigm shift. He has brilliant ideas, a cosmic perspective, and a holistic and dynamic worldview that far surpassed the science of his time and was not appreciated by his contemporaries."*

In the years since Capra wrote those words, Reich's orgone physics has remained largely uninvestigated, but it is of increasing interest to a new generation. From approximately 1947 to 1957, Reich laid the foundation for a new paradigm in physics, a new cosmology, and a new understanding of the nature of physical reality.

In *Ether, God and Devil* (1949), Reich described his thought technique—orgonomic functionalism—and revealed how its inner logic led him to the discovery of the cosmic orgone energy. Reich explored the concepts of "God" and "Ether," and compared and contrasted human thinking on these two profound concepts with his own scientific findings about orgone energy. His research to this point led Reich to conclude, in contradistinction to the physics of his day, that:

> There is no such thing as "empty space." There exists no "vacuum." Space reveals definite physical qualities. These qualities can be observed and demonstrated; some can be reproduced experimentally and controlled. It is a well-defined energy that is responsible for the physical qualities of space. This energy has been termed "cosmic orgone energy."

Today's space scientists might understand the above statement by Reich. His contemporaries did not.

In *Cosmic Superimposition* (1951), Reich posited that the cosmic orgone energy was the common functioning principle (CFP) that rooted man in nature. He described the superimposition of two orgone energy systems and demonstrated the existence of the function of superimposition of energy streams throughout nature—in spiral galaxies, hurricanes, the aurora borealis, and in the genital embrace of living organisms. In this book, he laid the foundation for a new cosmology.

In Reich's physics and cosmology, energy, and not matter, is the fundamental reality. Energy functions are involved throughout nature in the microscopic and the macroscopic realms; in the natural processes of birth, growth, decline, and death that hold true for the smallest individual life forms as well as for entire galaxies; and in the mysteries of sensation, perception, and self-perception.

In *Cosmic Superimposition*, Reich explored the ability of man to think and to know what nature is and how it works through thinking. He expressed a profound insight into the depth of the human longing for knowledge:

> The quest for knowledge expresses desperate attempts, at times, on the part of the orgone energy within the living organism to comprehend itself, to become conscious of itself. And in understanding its own ways and means of being, it learns to understand the cosmic orgone energy ocean that surrounds the surging and searching emotions.
>
> Here we touch upon the greatest riddle of life, the function of self-perception and self-awareness.

In *Cosmic Superimposition*, Reich investigated the unitary energy functions that are manifest in both the cosmos and consciousness.

In *The Oranur Experiment, First Report (1947-51)* Reich asked a daring question—could orgone energy influence nuclear energy and provide an antidote to the nuclear destruction of living systems? The Oranur (Orgone Anti-Nuclear) Experiment was designed to find an answer to that question. What did Reich discover? He wrote:

> The factual results were much more complicated than had been anticipated at the start. The dramatic, deadly, dangerous events during the Oranur experiment may easily obfuscate the final result which was positive. OR energy contains powerful functions directed against NR sickness, possibly even the power of immunization.

On February 3, 1951, three weeks after Reich exposed nuclear material to concentrated orgone energy during the Oranur Experiment, the *New York Times* reported that inordinately high background radiation counts had been recorded, in an area with a radius of three to six hundred miles. Rangeley was at the approximate center. These findings paralleled Reich's.

In addition, the Oranur experiment greatly advanced Reich's energy medicine. He discovered that medical Oranur effects "bring to the foreground specific disease characteristics of the individual." Reich had discovered a promising medically active agent that could search out specific disease syndromes and their exact locations in the body. With such a potent medical tool, the prevention of disease could become a reality.

In *Contact with Space* (1957), his last book, Reich reported on his weather modification work with cloudbusters in Tucson, Arizona; the antinuclear effects of orgone energy; the intimate connections among the spread of deserts on the planet, the "emotional desert" within human beings, and the presence of UFOs on Earth; and advanced mathematical gravity and antigravity equations that bear on the phenomenon of extraterrestrial spaceships and space travel.

An exciting area of investigation has now opened up—the relationship between Reich's orgone energy research from the 1940s and 1950s and present-day investigations of science, as illustrated by the phenomenon of zero-point energy fields.

- What is the relationship between the ubiquitous, limitless, primordial orgone energy ocean investigated by Reich and the ubiquitous, limitless zero-point energy sea of physics today?
- What is the relationship between the Casimir Effect and Reich's objective demonstration of mass-free orgone energy with the orgone energy field meter?
- What is the relationship between the so-called "vacuum energy" and the orgone energy Reich photographed and filmed luminating in a 0.5 micron pressure vacuum tube?
- What is the relationship of the motor force Reich demonstrated with orgone energy and the investigations of zero-point energy as a new source of power?
- What is the relationship between Reich's orgone energy research and the search of physicists for a unified field theory?
- What is the relationship between Reich's physics and cosmology and the development of scientific cosmology today?

There are many other questions, rich with potential, that may be asked as Reich's science is reevaluated in light of developments since his death in 1957. But, with this background on Reich's life and thought, it is now possible to turn attention to his work relating to the UFO phenomenon and extraterrestrial life. As with all of Reich's work, the orgone energy functions are the focus of investigation, the "red thread" that runs through the body of work.

## Reich, UFO Phenomena, and ET Presence

Reich did not like the accidental term *flying saucers*. He recommended changing the name flying saucers to "space ships." Reich referred to the occupants of the space ships as "CORE men," indicating that these spacemen were using the techniques of cosmic orgone engineering (CORE). As did most people who accepted the reality of the UFO phenomenon, Reich wondered why they were here. He asked, "What may have induced the CORE men to inspect so frequently the earth globe? . . . To know, really, we would have to meet them and ask them point blank."

Reich later proposed using the term *Ea* as an abbreviation for the visitors from outer space. He wrote that the *E* stands for "Energy" and the *a* for "alpha," or "primordial." At times, Reich wrote, *Ea* also stood for *Enigma*. The subject of extraterrestrial life, spaceships, and visitors from other solar systems remains an enigma today. But it may be an enigma that is about to be solved.

The problem of visitors from outer space did not draw Reich's attention until November 1953, when he read two books, *Flying Saucers from Outer Space* by Major Donald Keyhoe and *Flying Saucers Have Landed* by Desmond Leslie and George Adamski. In 1954, Reich read Keyhoe's book *Flying Saucers Are Real* and Frank Scully's *Behind the Flying Saucers*. The 1956 book, *The Report on Unidentified Flying Objects*, by E.J. Ruppelt, also influenced Reich's thinking.

One year earlier, in Maine in August 1952, Reich heard something speed from horizon to horizon in a few seconds. He said he knew immediately that it was a spaceship, but was not startled. Reich had long assumed that the universe was inhabited by intelligent beings, and so it did not surprise him that some might be exploring our planet.

After he read the early reports about the spaceships, Reich made a number of connections between the observations of those who saw UFOs and his observations of orgone energy functions. For example, the bluish lights associated with many reports of spaceships correlated with the blue color of orgone energy. In addition, the comparatively silent operation of the spaceships was compatible with the silent functioning of the orgone energy in the living organism and in the orgone motor.

Between 1940 and 1944, Reich had worked on advanced mathematical gravity and antigravity equations that were related to the phenomenon of spaceships and space travel. The tremendous speeds that were reported by those who had seen, photographed, or made motion pictures of UFOs were consistent with the calculations Reich had made a decade earlier. To Reich, speeds of 10,000 or 15,000 miles per hour would be quite natural for craft using orgone energy.

Reich felt that the CORE men operated their spaceships with the help of the cosmic orgone energy fields in space and around our planet. He believed the fantastic maneuverability of the spaceships, as described by eyewitnesses, indicated that the CORE men knew how to manipulate the planetary orgone energy fields that govern the function of gravitation. In addition, Reich speculated that the spaceships traveled in outer space on the Galactic and Equatorial orgone energy streams of the universe.

The rotating discs that were frequently reported in connection with UFO sightings also fit with Reich's orgone energy observations. "Rotating discs describe exactly what I had calculated 10 years previously . . . without any knowledge of space ships actually riding cosmic OR waves," Reich wrote.

Reich believed that orgone energy filled the universe and that the increased number of atomic explosions during the 1950s were irritating the highly sensitive cosmic orgone energy ocean. In the early 1950s, a majority of UFO sightings were at or near atomic installations, either military or civilian. As a result, Reich postulated that it was human atomic activity that attracted the attention of the CORE men. He wrote, "So far, to judge from the available reports, these CORE men came peacefully, investigating *only*."

Reich wondered if the intentions of the visitors from outer space were benign or hostile. He felt that the answer would be found in due time, but that it remained an open question. For

Reich, the visitors from outer space, and his orgone energy research, brought human beings into direct contact with space as it really was—and this was radically and fundamentally different from how it was conceived by the science of his day. Einstein's universe was an empty vacuum and we were alone; Reich's universe was filled with a pulsating cosmic energy and inhabited by living, intelligent beings.

The observations that Reich made and published about spaceships in the 1950s have been reported and documented since by others. According to Peter Robbins, author of a two-part article entitled "Wilhelm Reich and UFOs," published in 1990–1991 in the *Journal of Orgonomy*, "Literally every type of UFO-related phenomena Reich observed and recorded is documented in military and intelligence papers of the period or earlier. Many of Reich's conclusions about UFOs were shared or seriously considered, on paper, by highly placed individuals in military, intelligence, and government."

Robbins found that thousands of documents released under the Freedom of Information Act indicated that Air Force personnel, members of Air Force Intelligence, several astronauts, and two secretary generals of the United Nations had all made observations that were in line with Reich's. Also, documents released by the CIA, FBI, NSA (National Security Agency), AEC (Atomic Energy Commission), the Joint Chiefs of Staff, the Rand Corporation, and the Soviet National Academy of Sciences also contained information on spaceships similar to what Reich reported.

In *Contact with Space*, Reich published three photographs that are without peer in the annals of the scientific investigation of spaceships. They are brilliant examples of the application of the scientific method to the study of an unknown phenomenon. In fact, they may prove to be among the most important photographs in the history of ufology.

In Figure 1, Reich used a Leica camera with the aperture set at 3.5 and exposed the film for twenty minutes on a moonless night to get the "star tracks" photograph reproduced here. The camera was pointed toward a group of stars near the zenith. At first, the photograph may appear to be a quite ordinary picture. However, one "star" track in the photo (indicated by an arrow in the upper left corner) is longer than all the other tracks, which are of identical length. This should not be.

Reich examined the photograph even more closely and discovered that this same "star" track deviates from the parallel run of the rest of the star tracks by about 2.5 degrees. This, too, should not be. To Reich, these facts were startling and astounding. However, he declined to interpret them immediately.

Figure 2 and Figure 3 are two photographs taken by Reich on two consecutive clear nights in Arizona during the winter months of 1954–1955. He believed they showed "extraordinary celestial phenomena." The time of night, duration of exposure,

*Figure 1 (above): Star tracks. Exposure twenty minutes; full aperature 3.5; moonless night; track with arrow is longer than all the others; parallel lines; incidental finding: the longer track deviates by 2.5 degrees.*

*Figure 2 (page 280, top): An Ea under ORUR to the left (east) of trees. Note disruptions—"fading out."*

*Figure 3 (page 280, bottom): Ea, appearing suddenly (middle arrow, to the right of trees the following night at the same time and same exposure). Two upper arrows mark deviation and three times "fading out."*

**Mindshift in Science**

and the aperture and position were the same for both photographs. The distant brightness in the pictures comes from the lights of Tucson, eight miles south of Reich's temporary home in Arizona. These two photos show the tracks of *Ea*—or UFOs—in the night sky.

Figure 2 shows the track of the *Ea* low above the trees to the left (or east). In Figure 3 the same object is seen to the right (or west) of the same tree. Reich noted that the track in Figure 2 is only about half as long as the track in Figure 3 during the same period of exposure. And the track in Figure 3 appears suddenly in the picture, as if it came out of nowhere. Although the track in Figure 2 has several obvious interruptions, Reich writes that the object was still present in the dark areas. However, it did not luminate. In contrast, the track in Figure 3 shows no such interruptions in the emission of light. It appears suddenly (see the arrow in the center of the picture). Another object is seen above the track in Figure 3, apparently moving much faster. This second object is present from the beginning of the exposure and shows a "wobbling" curve (see first arrow) and three regular interruptions of light emission (see third arrow). In addition, there are other faint, barely visible tracks on Figure 3 that are not present on Figure 2.

Reich did not immediately interpret these pictures because he sought more information before he felt comfortable doing so. But he later concluded that these photographs suggest that all "stars" in the night sky may not be stars or planets. They may be space machines of extraterrestrial origin. In fact, during his research in Arizona, using the cloudbuster, Reich withdrew atmospheric energy from the region in which he saw two UFOs or *Eas*. The two "stars" to the west faded out several times when cosmic energy was drawn from their location. Reich did not publish these photographs to convince anyone of anything. The photos are examples of the scientific methods he used in his *Ea* research.

There is far too much material to present here about Reich's study of the UFO phenomenon. It is sufficient to note the substantive scientific contribution he made. Those interested in pursuing his investigations concerning spaceships and extraterrestrial life in detail will find a wealth of material in *Contact with Space* and in "Space Ships, DOR, and Drought," an article in Vol. VI, Nos. 1–4 of the journal *CORE* (Cosmic Orgone Engineering).

Did Reich prove that spaceships are visiting our planet in craft guided by intelligent beings? In *Contact with Space*, he asked "What do they want for proof? There is no proof. There are no authorities whatever. No President, no Academy, no Court of Law, Congress or Senate on this earth has the knowledge or power to decide what will be the knowledge of tomorrow. There is no use in trying to prove something to somebody who is ignorant of the unknown, or fearful of its threatening power. Only the good, old rules of learning will eventually bring about understanding of what has invaded our earthly existence."

On the threshold of the Cosmic Age, Reich was stopped by powerful social forces that converged to destroy both him and his work. In a statement included in his "Brief to the United States Court of Appeals for the First Circuit, October, 1956," Reich outlined his estimation of humanity's situation as it approached a "cosmic crossroads." He made the following assertions:

- The biological revolution of mankind is underway and cannot be stopped.
- Humans will develop a cosmic energy technology now known only to living beings from outer space.
- The *Cosmic Energy Motor* will be the lever which will turn our present civilization into that of the coming Cosmic Age.
- The cosmic OR motor will replace the motors of today and carry human space ships into the far reaches of the universe.

When one looks at Reich's accomplishments from 1923 to 1957, from his first investigations of the energy of instinctual drives to his explorations of orgone energy in the cosmos, one can only wonder how much more he would have discovered if powerful political, economic, and social forces had not successfully worked behind the scenes and used the U.S. court system to kill him and *temporarily* derail his science.

It is difficult to do more than speculate as to who was behind the effort to stop Reich. He believed Communists were involved in the attacks on him. The *New Republic* did have strong leftist, if not Stalinist, leanings and published the slanders against Reich that led to the FDA persecution. However, Reich's fascist enemies from Europe were also active and influential in America. The CIA's Operation Paperclip brought hundreds of Nazi psychiatrists to the U.S., many of whom joined the American Psychiatric Association, which actively worked with the FDA against Reich. In *American Swastika* and *Trading with the Enemy*, author Charles Higham has detailed the involvement of international fascism in the American pharmaceutical and energy industries, both of which were active against Reich and would have suffered financially if Reich's orgone energy had been proven real and been developed as a new energy source and new medical treatment.

Can we apply the Mindshift Hypothesis here? Were other forces at play behind the scenes? In addition to Reich's political enemies, was the clandestine group of which Edgar Mitchell speaks involved? Stanton Friedman's MJ-12 was formed just as Reich's mass-free energy physics was being developed. Did it play a role in Reich's persecution? Were his science and his mind too much for them to control? Did they consider his work a danger to "national security" and/or their own power base?

There is evidence indicating that Reich's work was known to some at the highest levels of the U.S. government and taken seriously. His energy discoveries *were* a challenge to individual character structure, powerful economic forces, the emerging National Security State—in other words, to the entire authoritarian social order itself.

Is it possible that the ETs themselves were involved in Reich's fate? That may strike some as a strange question indeed. However, anything is possible. In *Contact with Space*, Reich wrote that his Oranur operations seemed to be under careful observation by living beings from outer space. Reich also gave serious consideration to the possibility that a man who had worked with him on the orgone energy motor—and who had disappeared suddenly in 1949—may have been connected somehow with actual spaceships. But the truth is yet to be uncovered.

Reich considered himself to be a Planetary Citizen. He believed the species mankind is facing an emergency. He also believed that the principle of Life itself on Earth is being challenged. Reich wrote, "*We are in a process of deep and crucial change in our total existence,* biological, physical, emotional and cosmic."

What would visiting ETs see if they looked at our world? They would see that the human animal is facing an emergency. They would see humanity struggling for its life in a period of biological revolution. It would be obvious to the ETs that the human animal has lost its sense of connection with Nature and, as a result, is endangering all life on the planet. Present human activity to exploit Nature is causing the extinction of species on a stunning scale. The extinction rates for birds and mammals, for example, are one hundred times higher than the natural rate of extinction. Across the planet, frogs are giving birth to deformed offspring and untold numbers of species are dying off rapidly. Frogs survived whatever phenomenon killed off the dinosaurs, but they are not surviving the onslaught of the human animal. In the United States, nearly one-third of all plant species are in danger of extinction. Destructive human activity is so extreme that mass extinction could occur rapidly, within just one generation.

In nonliving Nature, visiting ETs would see that manmade chemicals have created an ozone hole over Antarctica that is twice the size of Europe. They would see that the resulting increase in ultraviolet radiation was reducing crop yields; causing an increase in cataracts and skin cancer; and suppressing the immune system of the human animal. ETs would certainly notice the desertification and deforestation of the planet; the climatic changes evident in the devastating droughts, floods, and storms of recent years; the shrinking polar ice shelves and melting glaciers; and the global warming and rising pollution of the atmosphere.

In observing human society, ETs would see the continuing carnage of warfare; the cruel abuse of children in myriad ways; the economic systems that value profit above all else. In studying human character structure, they would see a biologically rigid, armored human animal turned against itself and its own best interests. In other words, extraterrestrials would see us as we are—not as we wish to be, or fancy ourselves to be.

This is a time of crisis, and human existence does seem to be on the line. However, it is also a time of unparalleled opportunity. The tools humanity needs to do the work are at hand; the necessary knowledge is present. Will humanity rise to the challenge it faces? Can the power of love reassert itself in the people of Earth? Perhaps the words of the alien in the film *The Day the Earth Stood Still* are more true than we know, "Your choice is simple. Join us and live in peace or pursue your present course and face obliteration. We shall be waiting your answer. The decision rests with you."

If humanity is to survive, and thrive, it must look at itself honestly, and see itself as it really is. Illusions will destroy the human race. Deep contact with reality—no matter how painful—is essential if the human species is to survive. The problems the human animal faces are serious, but they are not insurmountable. However, the solution to the present crisis will involve the greatest effort yet made by humanity in its history. As we look at the mindshift underway in science; at the gradual bridging of the gap between space science and ufology; and at the renewed interest in orgone energy research; Reich's thoughts, published nearly fifty years ago in *The Oranur Experiment, First Report (1947–1951)*, are more pertinent than ever.

> *All boundaries between science and religion, science and art, objective and subjective, quantity and quality, physics and psychology, astronomy and religion, God and ether, are irrevocably breaking down, being replaced by a common functioning principle (CFP) of all nature which branches out into the various kinds of human experience.*

A new evaluation of the work of this pioneer is critical. It is well past the time that Reich's research be subjected to scientific scrutiny so that his revolutionary discoveries can be used to help create a better world. The development of Reich's orgone energy research could bring about the greatest mindshift of all.

*This spiral galaxy resembles a UFO because the cosmic energy that shaped the galaxy is the energy used by UFOs. Cosmic form follows cosmic energy function.*

PROJECT MINDSHIFT

# CHAPTER TWELVE

# The Alien Agenda
The Alien Agenda
The Alien Agenda
The Alien Agenda
The Alien Agenda

The last questions to be asked are the same as the first questions: Why are these extraterrestrial intelligences here? What is their agenda? Do they even have an agenda? Do different species of aliens have different reasons for being here? Are there even answers to these and similar questions?

There are tentative answers but it must be stressed that—in the public domain—no one has any definitive answers. In the domain of the clandestine groups that are believed to exist, there may be answers to these questions. But this information is kept from all but a few people, for reasons unknown.

Yet, one cannot help but ask, Why?

In Dr. Edgar Mitchell's view, there is an ET agenda but we do not know what it is. He acknowledges that those who claim to have been abducted by aliens report information that could be—and is—construed by some to accurately communicate what the alien agenda is. However, Dr. Mitchell himself does not know if the information offered by the abductees is credible. In addition, he does not think that evidence of an ET agenda is being kept from the public in any systematic way.

To Budd Hopkins, the alien agenda seems clear. The UFO and alien abduction phenomena are leading to the creation of alien-human hybrids here on Earth. Hopkins believes that the alien study of humanity and use of human beings—almost as lab animals—focuses on genetics and reproduction because these factors are central to the alien hybrid program. He does not view this alien activity as necessarily beneficial to human beings. In addition, Hopkins feels that this program is not new. He has hundreds of cases that date back to the 1930s and 1940s. In fact, Hopkins was the first ufologist to deal with this possibility, doing so in his book *Intruders*. The question remains unanswered as to why the aliens wish to create hybrid beings.

David Jacobs, Ph.D., believes that he has a possible answer as to why the aliens are engaged in a massive abduction-hybrid program. Based on what he sees as reliable evidence gathered from hundreds of human beings who have had encounters with aliens, it is his opinion that the alien activity is in the service of a goal-directed breeding plan designed to create human-alien hybrids, a kind of *Homo Alienus*, who will then be integrated into human society. These hybrids will have a task and a "mission" on Earth and will play key roles in restructuring human society. In addition, according to Dr. Jacobs, the hybrids will mate with human beings. They will be indistinguishable from normal Homo sapiens, except for one characteristic—*their biological energy field.* Those with the eyes to see will be able to distinguish the hybrid beings from ordinary Earthlings. For Dr. Jacobs, as for Budd Hopkins, this alien activity may not be beneficial to human beings.

Ufologist Raymond Fowler has also given thought to the problem of the purpose and intent of the alien activity on Earth. He said, "Yes, there is an alien agenda—but what is it? My dog might see me out in my garden, digging. From his point of view, he probably thinks I'm looking for a bone. Each one of us is looking at one facet or two facets of the UFO phenomenon and saying, 'I like that! It seems to fit.' Things that don't fit, we ignore. The 'alien agenda' is built on what seems acceptable to the researcher, and everything else is rejected."

Dr. John Mack has not been able to conclude from his work with abductees or, rather, experiencers, that there are discrete, different types of alien beings with either an agenda or a number of agendas. In fact, he is not even comfortable with the word "agenda." He views the UFO phenomenon as part of an evolutionary process that is inextricably linked with the problems of existence, consciousness, and self-awareness. There may be an "agenda," but Dr. Mack asked, "Whose agenda is it? Is it the aliens' agenda? Is it our agenda? Or is it God's agenda, or the Cosmos's own agenda, depending on whether or not one has a deistic or nondeistic theology?"

For Don Berliner, it is impossible to understand alien behavior because human beings are completely ignorant of the psychology of intelligent non-human life. Human beings are also completely in the dark about the structure and nature of alien civilizations and cultures. In his opinion, any theories about an alien agenda can only be based on anthropomorphic projection, and, therefore, of little value.

Futurist and journalist Michael Lindemann has looked at this difficult question as well. He observed, "If there are any aliens here at all, we would have to rationally surmise that there are several different kinds. If that is true, then there would be several different agendas. There is no particular reason to believe that these agendas are all coherent and cooperative. To be frank, I don't see anything I could call an alien agenda. I see many, many instances of apparent contact between humans and other intelligences. That is what it looks like to me. I do not see the hard-edged pattern that David Jacobs is so sure that he can see, as described in his book *The Threat*. Budd Hopkins and David Jacobs essentially see eye-to-eye on this. I'm not saying they are wrong. But I am saying that I don't see the pattern the way they see it."

Lindemann pointed out a few different theories concerning the alien agenda, such as the "lab animal" theory, the "Space Brothers" theory, and the "Satan's Minions" theory. "Over the next few years we will see the Satan's Minion theory about aliens grow. It will become an increasingly vituperative part of the UFO conversation. For example, Pat Robertson recently said that people who believe in UFOs should be stoned because they are, in effect, colluding with Satan. There is going to be more of this kind of religious fascism. On the other side, you have the Space Brothers theory, of which there are many variations. Donald Ware, a very smart man—a lieutenant colonel and nuclear engineer—has concluded in the last few years that there is a very deep relationship between the aliens and the U.S. government. He believes the government is working to move us to a time when we will all be welcomed into the Cosmic Federation and everything will be great."

Lindemann expressed reservations about such conclusions. As a journalist, he attempts to be a fair broker of a wide range of UFO material, no matter what his personal opinions might be on any particular viewpoint. In fact, the most important aspect of the UFO dialogue underway may not turn out to be which view or opinion is "correct." The most important aspect may be the UFO conversation itself. To Lindemann, "Any particular take on UFOs is just a piece of the puzzle. History is the product of a conversation that is constantly occurring. And this conversation is very much a part of the history-making process today. But, to me, no particular story or theory about aliens—no particular voice or strain—stands out. *The real story is the cultural ferment,* not any particular take on UFOs."

Stanton Friedman, in contrast to most other UFO investigators, states clearly exactly what he thinks the alien agenda is. "I think that their primary agenda is to make sure that we don't get out there until we get our act together. In my paper, 'A Scientific Approach to Flying Saucer Behavior,' I present a list of twenty-six reasons for coming to Earth: ETs could be graduate students doing their theses on the development of a primitive society; they could be broadcasters doing a weekly show; or they could be mining engineers exploring the Earth, the densest planet in the solar system, for heavy metals. There could be a zillion different reasons the aliens would come here.

"But—I make one important assumption about every advanced civilization. Namely, that it is concerned with its own survival and safety. That seems to me to be a reasonable assumption. That being the case, I think that at the end of World War II, the 'alien cop on the beat' would have been very concerned about this primitive society whose major activity is tribal warfare. We provided three signs or indications that soon—within one hundred years or less— these idiot Earthlings would be bothering them. One hundred years is not much on the Cosmic Timescale. We would be taking our particular brand of friendship—*hostility*—out there. After all, we had just killed fifty or sixty million of our own kind. We had just destroyed seventeen hundred cities in the world war."

According to Friedman, the three visible signs or manifestations of impending human space travel were the development of: one, atomic bombs, which contain a great deal of energy in a small package; two, powerful V-2 rockets; and three, powerful new radar technology, whose signal goes out into space. In his opinion, when put together, these three developments make it clear that human beings would be traveling to deep space within a century.

Friedman noted, "If you were an alien, would you want Earthlings out there? There are a lot of good things here on Earth but every single day of the year, *thirty-five thousand children die needlessly of preventable disease and starvation.* While we continue to spend *three-quarters of a trillion dollars annually* on things military."

How does Friedman—one of the few scientists in the field of ufology—see humanity's place in the Universe? "I think we are the Johnny-Come-Latelies in the neighborhood. After all, Zeta 1 and Zeta 2 Reticuli are only about *three light weeks* from each other. We are about *4.5 light years* from the nearest star." This is highly significant because the inhabitants of Zeta 1 and Zeta 2 Reticuli, when they reached the stage humans did, with the development of the telescope, could directly observe each other's planets all day long. They would have been able to see signs of life. *They would have known they were not alone.*

Friedman continued, "I would expect there to be earlier extrastellar travel when your nearest neighbor is only three light weeks away. Zeta 1 and Zeta 2 are only thirty-seven light years from us— *and a billion years older than us.* I suspect there are a lot of civilizations in the neighborhood. I wouldn't blame any of them for making sure these idiot Earthlings—who obviously don't have their act together—don't come out there. I can't imagine the Galactic Federation allowing representatives from individual nations, any more than the United Nations allows cities to join."

If there were an alien agenda, and if the intelligent extraterrestrial entities visiting Earth wished to speak to Earthlings about their plans or concerns, to whom would they speak? Friedman asks a simple question to which there is no simple answer: "*Who speaks for Planet Earth?*"

If living beings from other worlds traveled through space, came upon Planet Earth, and studied the human animals that live here, what would they see? Astute alien observers would recognize that the human animal behaved as if it had lost its sense of connection to Nature. The degradation of the environment that is underway—the great "ecological holocaust" as some call it—would be apparent even from space. Certainly, extraterrestrials would also notice the global outbreaks of "tribal warfare" that Stanton Friedman describes as our primary activity.

A closer look at human society would reveal that humans were making some progress in their own efforts to leave the surface of Earth and explore space. Extraterrestrial observers would also see that humans had developed powerful but primitive technology, as well as various forms of social organization to ensure survival and manage the interaction and behavior of large groups of people. In addition, the extraterrestrials could not help but notice that the human animal had developed the capacity to express itself through science, religion, the arts, education, etc.

Yet, it would be obvious to an observer from another world that—despite all of humanity's accomplishments—something was fundamentally wrong. If an alien civilization familiarized itself with the history of human thought, it would discover that mankind has searched in vain through its entire written history for an answer to the question, *What is wrong with us?*

Plato compared humanity to people living in a cave who had adapted to the dark. Cut off from contact with the world outside of the cave, they had learned to read the shadows on the walls. Over time, these people mistook the shadows on the wall for the real things. In addition, they had grown to hate the light. Having lived so long in darkness, they could not tolerate the light of Life.

Moses sought to lead humanity from slavery that seemed to be imposed from without. Yet after the escape from Egypt, the inner chaos of the people emerged. They had become incapable of living in freedom. In an attempt to keep a conflicted and confused society from crumbling altogether, a rigid code of "thou shalt nots" was developed and imposed from without. As once behavior had been determined by the force of the wrathful and vengeful pharaohs, now it was enforced by a strict moral code backed by the myth of a wrathful, vengeful deity.

In another attempt to save humanity from itself, Jesus brought a message of love into the world during the rule of the cruel Roman Empire. Humanity loved his beautiful dream but lacked the capacity to make it real. The people did not have the ears to hear or the eyes to see. Hope turned to hatred and Jesus died, crucified in ignominy. Christ's heirs twisted his simple, cosmic message of love for one another—expressed through the body, heart, and soul—into its opposite: an ascetic, antisexual, judgmental, life-negative theology imposed through power and maintained through fear.

There were many other efforts to describe what is wrong with humanity, and many examples of other great attempts to change people and restructure society. In the twentieth century, Freud's revolutionary science was turned into a business and Marx's utopian vision of international brotherhood was perverted into totalitarian, power-political, national states.

The great question—*what ails mankind?*—remains unanswered.

If intelligent alien beings were studying humanity, is it not likely that they too would seek to find the answer to that question, an answer that has eluded mankind for thousands of years? If so, where might they look? It is quite possible that in studying the human animal, an alien culture might begin at the beginning—*gestation*, the period of development in the uterus from conception until birth. Perhaps, they might reason (if that word is even applicable) that the answer to what plagues the human animal may begin in the uterine environment. Or the problems may stem from events surrounding birth or which occur in the immediate period after birth.

What would aliens find if they examined how human beings treat their newborn infants? They most likely would conclude that, in general, all across Earth, for reasons yet to be discovered, the human animal treats its newborn infants cruelly. The infant is slapped; the umbilical cord is cut, instead of allowed to wither naturally, as the infant begins to breathe on its own. The infant is wrapped in tight garments, or even swaddling clothes, and is separated from the mother. The newborn is often subjected to harsh chemicals and drugs. In many cultures, the infant reaches out for the mother's breast, only to be given a bottle with a cold rubber nipple. Or the mother attempts to breastfeed the child, but the nipple won't get erect, or the milk won't come, or the milk is sour. This occurs under normal circumstances, where people believe they are doing their best. The widespread abuse and neglect of infants is another tragic reality that would be apparent to any intelligent non-human studying humanity.

In the United States, this common mistreatment of infants is often followed by a cruel "routine" surgical procedure involving the male infant's genitals; namely, routine circumcision.

Circumcision is regularly performed on infants without anesthesia. What other surgical procedures do parents allow to be performed on their infants without anesthesia? Religious, medical, hygienic, and cultural rationales are offered to justify this cruelty to infants. Observers from other worlds may look at the impact of the procedure on the infant and base any judgments on that, and not on theology or ideology.

In the United States today, there is vocal criticism of "female genital mutilation" as it is practiced in other cultures. Is "routine circumcision" *essentially* any different from female genital mutilation, or do the procedures differ only in degree? The extraterrestrial observers may see obvious links between routine circumcision and other forms of circumcision. To an observer from another world, the common feature—an invasive surgical procedure involving the genitals—may be more significant than any differences because of the deleterious consequences to the newborn.

When an infant is born a new bit of the cosmos, organized in the form of intelligent life, enters our world. This infant—this living bioenergetic system—reaches out to its environment and, in general, it meets a harsh, callous world, and is rejected. The infant then gradually stops reaching out toward the world and pulls back, away from the world. It is here—in the mistreatment of the newborn—that so many physical, emotional, psychological, and social problems have their origins.

Can we give consideration to the possibility that intelligent non-human lifeforms may look at the human animal today and see a species that, from birth onward, interferes with the natural development of the sexuality of its offspring? Might not alien beings notice that this interference has serious consequences? There is a vast body of medical, psychological, anthropological, cross-cultural, and even literary evidence indicating that sexual disturbances affect many aspects of life. The study of human sexuality would certainly be a central focus of any extraterrestrial investigation of the human animal.

Before we get too far afield, let us ask this question: How may the subject of human sexuality relate to the UFO phenomenon? It is related in terms of *energy*. In the ancient Taoist view, there was one primoridal energy manifest in all of Nature—in the cosmos; in the atmosphere of our planet; and in all living organisms. In the twentieth century, Reich pursued energy investigations, showing that Freud's *psychic energy* was actually *sexual energy, a specific bioenergy.* Further experimentation indicated that this bioenergy was a function of an *atmospheric energy* that was itself a manifestation of a *cosmic energy.* The ancient insights seem to be confirmed by contemporary scientific investigation. There appears to be *one primoridal cosmic energy* at work in the Universe, expressing itself as various functions in the living and nonliving realms. If this is so, surely extraterrestrial space travelers are aware of this cosmic energy. It may be this primordial energy that the aliens are studying in people.

Budd Hopkins and David Jacobs believe they have evidence from abductees of an intense alien interest in human sexuality. Others who have worked on the UFO abduction phenomenon have reported this alien interest in human sexuality as well. To Hopkins, Jacobs, and others, the aliens seem to be carrying out a massive, covert program whose focus is on human genetics and reproduction. They are convinced of this, and both researchers believe that they have the evidence to back up their views.

However, knowledgeable people such as Edgar Mitchell have stated that it is not possible to know with certainty if abductee information is credible. As Don Berliner has pointed out, we know that the memories of abductees have been altered. However, we do not know how they have been altered. For example, if the aliens can use mind control techniques to cause amnesia in abductees, is it not at least a serious possibility that the hypnotically retrieved memories of the abductees might actually be false memories, put there by the aliens themselves, as a cover? There is no way to really know.

As convincing as the abduction stories are—and as honest and credible as the abductees themselves are—it is still impossible today to judge with certainty how literally to take the abduction reports. We do not have that problem with human behavior, however. Human sexual dysfunction is evident in nearly every newspaper or magazine, on radio and television, in movies, and on the Internet. This surely cannot escape the notice of alien beings who may be monitoring our mass media.

It is worth considering that human sexual function may interest the aliens because it plays a key role in what ails mankind. There is ample evidence to link sexual pathology with a wide range of destructive and self-destructive behavior. The ability to work, think clearly, act independently, feel pleasure, and sense a connection with Nature are all impaired when the sexual function is impaired. In addition, new disciplines such as psychoneuroimmunology (PNI) show the link between sexuality, emotions, health, and disease. Perhaps ET interest in human sexuality has more to do with this biological tragedy than we know.

Before humanity can connect with the Cosmic Community, it has to heal, on every level, including the sexual. Only then can human beings reconnect with Nature, inside and outside of themselves. Humans *are* rooted in Nature, although they have lost contact with this reality. The intelligent, extraterrestrial, non-human lifeforms are rooted in Nature as well. It is crucial to keep this in mind.

The story of the interaction between European explorers and the indigenous peoples of the New World can be instructive here. When the Europeans arrived in their sailing ships, many of the indigenous peoples interpreted the appearance of these newcomers in terms of their religious mythology. At first, they invested the Europeans with "superior" or even "godly" attributes on the basis of their sudden mysterious appearance and on their advanced technology.

To the indigenous peoples, the arrival of these unknown lifeforms seemed to have a "meaning" in relation to their belief systems. However, the Europeans came to the Americas for reasons of their own, which had nothing at all to do with the indigenous peoples. The religious-mystical interpretation of the arrival of Europeans on their shores contributed to the downfall of the existing cultures. The visitors were not gods or advanced beings; they were simply other flawed lifeforms in their own stage of evolution.

Similarly, the aliens may have come here for reasons that have nothing to do with humanity. Human beings ought to refrain from ascribing a meaning to their arrival. In addition, it is not wise to invest the unknown nonhuman intelligences with any superior or godly attributes, especially on the basis of the advanced technology they appear to posssess. Humanity has made astounding technological leaps in the twentieth century, but it has not advanced greatly on other levels of existence in thousands of years. Advanced entities may be studying us, but they are not necessarily inherently superior. Most likely they are merely other natural expressions of the Life Energy in the

Universe, exploring creation. Whatever they may be, no good can come from mistaking these entities for something they are not. This can only block true understanding of who they may be.

The alien agenda has been shrouded in mystery for fifty years, and may remain so for another fifty years. But the human agenda is clear. Shakespeare wrote that the fault is not in the stars, but in ourselves. So too, the answers are not in the stars or UFOs, but in ourselves. And the answer we need most is the oldest answer of all—*unconditional love*, the key to reconnecting with the cosmos. That is the human agenda, and it is time we got to it.

The hour is getting late.

# SELECTED READINGS

Berliner, Don. *Unidentified Flying Objects Briefing Document.* UFO Research Coalition, 1995.

———— and Stanton Friedman. *Crash at Corona: The U.S. Military Retrieval and Cover-Up of a Crashed UFO.* Paragon House, 1992.

Brown, Courtney. *Cosmic Voyage.* Dutton, 1996.

Bryan, C. D. B. *Close Encounters of the Fourth Kind.* Knopf, 1995.

Corso, Col. Philip J. *The Day after Roswell.* Pocket Books, 1997.

Drake, Frank and Dava Sobel. *Is Anyone Out There?* Delta, 1992.

Emenegger, Robert. *UFOs Past, Present and Future.* Ballantine, 1974.

Emmons, Charles E. *At the Threshold: UFOs, Science and the New Age.* Wild Flower Press, 1997.

Fawcett, Laurence and Barry Greenwood. *Clear Intent.* Prentice-Hall, 1984.

Fowler, Raymond. *The Andreasson Affair.* Prentice-Hall, 1979.

————. *The Watchers, Part I.* Bantam, 1990.

————. *The Allagash Abductions.* Wild Flower Press, 1993

————. *The Andreasson Affair, Phase Two.* Wild Flower Press, 1994.

————. *The Watchers, Part II.* Wild Flower Press, 1995.

Friedman, Stanton. *Top Secret/Majic.* Marlowe, 1996.

———— and Don Berliner. *Crash at Corona: The U.S. Military Retrieval and Cover-Up of a Crashed UFO.* Paragon House, 1992.

Good, Timothy. *Above Top Secret*. Morrow, 1988.

Hill, Paul R. *Unconventional Flying Objects: A Scientific Analysis*. Hampton Roads, 1995.

Hopkins, Budd. *Missing Time*. Ballantine, 1981.

———. *Intruders*. Ballantine, 1987.

———. *Witnessed*. Pocket Books, 1996.

Hynek, J. Allen. *The UFO Experience*. Ballentine, 1972.

Jacobs, David. *The UFO Controversy in America*. Signet, 1975.

———. *Secret Life*. Simon & Schuster, 1992.

———. *The Threat*. Publisher, 1997.

Lindemann, Michael. *UFOs and the Alien Presence: Six Viewpoints*. 2020 Group, 1991.

Mack, John E. *Abduction: Human Encounters with Aliens*. Scribners, 1994.

Marrs, Jim. *Alien Agenda*. HarperCollins, 1997.

Mitchell, Edgar. *The Way of the Explorer*. Putnam, 1996.

Peebles, Curtis. *Watch the Skies!* Berkley Books, 1995.

Pritchard, Andrea, David E. Pritchard, John E. Mack, Pam Kasey, and Claudia Yapp (eds.). *Alien Discussions: Proceedings of the Abduction Study Conference held at MIT*. North Cambridge Press, 1994.

Randles, Jenny and Peter Worthington. *Science and the UFOs*. Basil Blackwell, 1985.

Reich, Wilhelm. *Cosmic Superimposition*. Wilhelm Reich Foundation, 1951.

———. *The Oranur Experiment*. Wilhelm Reich Foundation, 1951.

———. "Space Ships, Drought and DOR." CORE, Vol. VI, Nos. 1–4, July 1954.

———. *Contact with Space*. Core Pilot Press, 1957.

———. *Ether, God and Devil*. Farrar, Straus and Giroux, 1973.

Robbins, Peter and Larry Warren. *Left at the East Gate: A First-Hand Accound of the Bentwaters-Woodbridge UFO Incident, Its Cover-Up and Investigation*. Marlowe, 1997.

Rux, Bruce. *Hollywood vs. the Aliens*. Frog, Ltd., 1997.

Sagan, Carl and Thornton Page. *UFOs: A Scientific Debate*. Barnes & Noble, 1996.

Sellier, Charles. *UFO*. Contemporary Books, 1997.

## UFO Organizations

**Center for UFO Studies**
2457 West Peterson Avenue
Chicago, IL 60659-4118
(312) 271-3611

**Fund for UFO Research**
P.O. Box 277
Mount Rainier, MD 20712
(703) 684-6032

**Mutual UFO Network**
103 Oldtowne Road
Seguin, TX 78155
(210) 379-9216

## Other Related Organizations

**Friends of the Institute of Noetic Sciences (FIONS)**
162 Fifth Avenue, Suite 1005
New York, NY 10010
(212) 741-2207
Fax: (212) 741-3794
Email: fions@mindspring.com
Website: http://www.fions.org

**Program for Extraordinary Research Experience (PEER)**
P.O. Box 390707
Cambridge, MA 02139-0008
(617) 497-2667
Fax: (617) 497-0122

**The Wilhelm Reich Museum**
P.O. Box 687
Rangeley, ME 04970
(207) 864-3443
Email: wreich@rangeley.org
Website: http://www.sometel.com/~wreich

**Intruders Foundation**
P.O. Box 30233
New York, NY 10011
(212) 645-5278

I would like to thank the following people for their contributions to this book: Tami Coyne for her close reading of the manuscript and astute observations, her friendship, love and support, and for helping me to "snap out of it" in times of difficulty and stress; Patricia B. Corbett, for the love, support, guidance, wise counsel and specific suggestions she offered; Debra Buck, for her excellent research work on this project, and for years of sharing her insights with me and helping me sharpen my own thinking; Robert Mannion, for his computer graphics work; Bill Behnken, for his wonderful UFO drawing on page 20; Katherine Sands, my agent, for making first contact with editor Betty Anne Crawford; and George C. de Kay, my publisher, Karlin Gray my editor, publicist Darcie Rowan and everyone at M. Evans and Company who worked on this book.

I would also like to thank my brothers George, Jerry, Bob, Tom and Jim, and sister Mary (for reasons I am sure will be revealed to me at a later date) and, in alphabetical order, my friends, Pat and Jerry Adler, Bill Behnken; Anthony Bernini; Rose Marie Bressan, Mary Ann Cleaves; Michelle and Brian Driscoll; Charles Flax; Mike and Mary Fletcher-Rogers; Michael and Barbara Gentilesco; Claire Gerus; Robert Gottlieb; Jan Kardys; Cynthia Kear; Gary Kniffen; Malka Margolies, Jean McCarthy, David Mendelsohn, Paul Peterson, Ruth and David Price-Chu; Ted Rado; my favorite cousin, Geraldine Romano; Bob Spencer, Martha Sturm, Kathleen Taggart; Don Tapken; Denise Trupia; Adam Tyler; and Fred and Patricia Zeserson, all of whom, no matter what their opinions may be on the content of this book, have helped me and shared their lives with me, in good times and bad. I hope I have not inadvertently left anyone out. If I have, count yourself in.

In addition, I want to thank all the researchers and investigators who took the time to be interviewed for this book: Don Berliner; Raymond Fowler; Stanton T. Friedman; Budd Hopkins; David Jacobs, Ph.D.; Michael Lindemann; John E. Mack, M.D.; Edgar Mitchell, Sc.D.; and Peter Robbins. My thanks to Jayne Allyson and Helen Wheels for allowing me to interview them and share their experiences with the reader. I am very grateful to John W. White for taking the time to review my manuscript and for writing a wonderful Foreword to the book.